Lecture Notes in Mathematics

Editors:
A. Dold, Heidelberg
F. Takens, Groningen

T0250599

Springer
Berlin
Heidelberg
New York
Barcelona
Budapest
Hong Kong
London
Milan
Paris
Santa Clara
Singapore
Tokyo

Jesús M.F. Castillo   Manuel González

Three-space Problems
in Banach Space Theory

Springer

Authors

Jesús M. F. Castillo
Departamento de Matemáticas
Universidad de Extremadura
Avda. de Elvas s/n
E-06071 Badajoz, Spain
E-mail: castillo@ba.unex.es

Manuel González
Departamento de Matemáticas
Universidad de Cantabria
Avda. de los Castros s/n
E-39071 Santander, Spain
E-mail: gonzalem@ccaix3.unican.es

Library of Congress Cataloging-in-Publication Data

Castillo, Jesús M. F.
    Three-space problems in Banach space theory / Jesús M.F. Castillo,
Manuel González.
        p.   cm. -- (Lecture notes in mathematics ; 1667)
    Includes bibliographical references (p.   -   ) and index.
    ISBN 3-540-63344-8 (softcover : alk. paper)
    1. Banach spaces. 2. Ultraproducts. 3. Duality theory
(Mathematics)   I. González, Manuel. II. Title. III. Series:
Lecture notes in mathematics (Springer-Verlag) ; 1667.
QA3.L28   no. 1667
[QA332.2]
510 s--dc21
[515'.732]

Mathematics Subject Classification (1991): 46B03, 46B20, 46B08, 46B10

ISSN 0075-8434
ISBN 3-540-63344-8 Springer-Verlag Berlin Heidelberg New York

© Springer-Verlag Berlin Heidelberg 1997
Printed in Germany

Typesetting: Camera-ready output (Word Perfect 5.1) by the authors
SPIN: 10553314   46/3142-543210 - Printed on acid-free paper

Foreword

A three-space problem, in the Banach space setting, has the form: Let P be a Banach space property, let Y be a subspace of X and let X/Y be the corresponding quotient space. Is it true that X has property P when Y and X/Y have property P ? If the answer is positive then P is said to be a *three-space property*. We shall often shorten "three-space" to $3SP$ (see below).

Three-space questions face the general problem of the structure of arbitrary subspaces and quotients of Banach spaces. Nowadays, it is clear that only Hilbert spaces can be labelled as simple: all subspaces are complemented Hilbert spaces and all quotients are Hilbert spaces. Other spaces may contain (and usually do) bizarre uncomplemented subspaces and/or quotients. The point of view of $3SP$ problems is to consider the structure of a Banach space contemplating *as a whole* the structure of subspaces and the quotients they produce. Nevertheless, as it is noted in [136], "it may happen that a Banach space with quite complicated structure may possess nice factors through nice subspaces." When a property turns out to be a $3SP$ property, it means that, despite the complexity that the combination of subspaces and quotients may have, the structure related with P is maintained. Incidentally, this gives a way to prove that a space possesses property P.

In the study of a $3SP$ question it is interesting, when the answer is affirmative, to uncover the structure behind the proof; for instance, *being finite-dimensional, reflexive,* or *no containing* l_1 are all $3SP$ properties for essentially the same reason: the possibility of a lifting for the different types of sequences considered (resp. convergent, weakly convergent or weakly Cauchy). When it is not, the construction of a counterexample is more often than not a rewarding task. Also, there is the general question of how to construct the "middle" space starting with two spaces having P: one space Y that plays the role of the subspace and other Z that plays the role of the quotient. The simplest form to do this is making their direct sum $Y \oplus Z$. In direct sums the factor spaces appear complemented. Spaces admitting Y as a non-necessarily complemented subspace and Z as the corresponding quotient are called "twisted sums." The interesting general problem of how to construct twisted sums is considered, at an elementary level, in Chapters 1 and 3. This problem spreads into many ramifications: categorical methods, semi L-summands, inductive limits, construction of $C(K)$ spaces, interpolation theory, etc, each of them contemplating a different aspect of the problem.

Conversely, a negative answer to a $3SP$ question is also of interest because it must involve a new construction of a space or an operator (usually the quotient map). It is then no surprise that some $3SP$ examples can be translated into examples of pathological operators; this topic has not been sufficiently exploited in the literature, reason for which we present an overall view in an appendix to Chapter 2.

The origin of $3SP$ problems can be traced back to a problem of Palais: If Y and X/Y are Hilbert spaces, has X to be isomorphic to a Hilbert space? Chronologically, $3SP$ results did not formally appear until middle seventies, although Krein and Smullian's proof that reflexivity is a $3SP$ property came around 1940. Counterexamples appeared with Enflo, Lindenstrauss and Pisier's solution to Palais problem (1975), while methods can be seriously considered after Jarchow's paper (1984). In this case, however, methods and counterexamples can be thought of as two different approaches to the same problem: some counterexamples provide a method, and methods sometimes provide a counterexample. Examples of counterexamples that provide methods could be the Kalton and Peck solution to Palais problem, which gave birth to the beautiful theory of quasi-linear maps; or Lusky's proofs that every Banach space X containing c_0 admits a subspace Y such that both Y and X/Y have a basis.

Currently it is a standard question to ask for the $3SP$ character of a new property. For this reason we devoted Chapter 2 to explain methods to obtain $3SP$ properties. These methods include (and, we hope, cast some light on) some classical topics such as lifting results or factorization of operators. The counterexamples can be found in chapters 3 to 7 collected in what could be seen as something like a zoo.

Thus, we propose the reader a guided tour through that zoo. These notes exhibit alive all (all?) specimens of $3SP$ problems that have been treated in the literature and which freely lived disseminated in many research papers, their natural habitat since they scarcely appeared in books. They have been patiently hunted and captured, carefully carried over (the common features of counterexamples have been emphasized), neatly polished (we often give simpler proofs) and classified following their nature (the proofs were unified through several general methods). There are also many new results and open problems.

The notation we follow is rather standard except in one point: the abbreviation $3SP$ for "three-space." Although it is not standard, it is clear, clean and direct; started with our e-mail messages and has achieved some success.

We have intended to give an essentially self-contained exposition of $3SP$ problems in the context of Banach spaces. There is a vast unexplored land of $3SP$ problems in other contexts (locally convex spaces, Banach algebras,...) that we shall not consider. All properties appearing in the book are defined and their basic relationships stated with appropriate references. Thus, some bounds about where to stop had to be imposed: things directly involved with $3SP$ questions appear in detail; other results shall be just cited when they have already appeared in books. The word **Theorem** is reserved for results having the form: property P is (or is not) a $3SP$ property. We made an effort to keep the reader informed about who did exactly what; this information is often given during the introductory comments although, in some cases, even conversations with the authors could not completely clarify where the ideas came from, and so we give our interpretation. About the problems and questions scattered through the text, we can only say that, to the best of our knowledge, they are open; some appear certainly to be difficult, some could even be easy, all seem interesting.

The background required to read these notes is a course in functional analysis and some familiarity with modern Banach space methods, as can be obtained from, e.g., J. Diestel's *Sequences and Series in Banach Spaces*. In order that the book can be used as a reference text we have included a summary, in alphabetical order, containing four entries for each property: 1) name; 2) yes, not, or open, for the corresponding $3SP$ problem; 3) general method of proof, counterexample or additional information; 4) location of the result in the book.

Many people provided us with a constant flow of preprints, information, questions or advice that kept the project alive. To all of them our gratitude flows back. Working friends Félix Cabello, Rafael Payá, Fernando Sánchez and David Yost wrote several new proofs for this book, tried to re-do some sections ... and we are sure they would have written the entire book ... had we left them the opportunity. Susan Dierolf, Ricardo García, Hans Jarchow, Mar Jiménez, Antonio Martínez-Abejón, Mikhail Ostrovskii and Cristina Pérez made useful suggestions. Moreover, thanks should go to the colleagues of the Mathematics Department of the Universidad de Extremadura and of the University of Cantabria for providing the natural conditions for working. The topic of the book was discussed with colleagues and lectured on at several meetings during these three long years of elaboration: the Curso de Verano of the Universidad de Cantabria en Laredo; the Analysis Seminar at the Universitá di Bologna, organized by Piero Papini; the Conference on Functional Analysis at Camigliatello, organized by Antonio Carbone and Giuseppe Marino; the Conference on Function Spaces at the Universidad Complutense de Madrid, organized by José Luis LLavona and Angeles Prieto; and the 2nd Congress on Banach spaces at Badajoz. Thanks are due in each case to the organizers and supporting institutions.

Beloved Boojums and the Bellman. The first author thanks the students of the Analysis seminar who attended a course on $3SP$ problems, and especially to F. Arranz "Curro" who arranged a place for some invention. During this time, what started as a harmless photographic safari became a frumious hunting, sometimes galumphing. Perhaps no one has described it better than Lewis Carroll in *The Hunting of the Snark*; we suggest a careful reading of it to anyone interested in mathematical research. For instance, take the last but one stanza of the third fit.

```
I engage with the Snark - every night after dark -
          In a dreamy delirious fight:
   I serve it with greens in those shadowy scenes,
          And use it for striking a light.
```

On a higher ground, it seems to me that my mother and my sister have been working almost as much as I did to carry this book through. Without their supporting hand and their endless patience I would probably have softly and suddenly vanished away. This not being so, the real $3SP$ problem for me has been then to combine the space of personal life, that of my University duties and the space required to make the book. Luna was the (precious) kernel that made this sequence exact. And one cannot leave unmentioned the exact sequence of our dogs: Suerte, Burbujas and Schwarz —which is my sister's dog, but anyway. Luna's mother and father, Javier Blanco, Antonio Hinchado (as the Banker) and Paco Pons (as the Booker) made a nice crew that helped to mix the bowsprit with the rudder sometimes.

The manuscript was patiently typed by Luna Blanco. Financial support came in part from the DGICYT (Spain) Project PB94-1052.

Although Bellman [51] affirms *"everything I say three times is true"* that does not necessarily apply to everything *we* say about three-spaces.

Contents

Chapter 1

Three-space constructions

This chapter is about the structure of a three-space (in short, $3SP$) construction. We first show the tools of the theory of categories that shall be needed: short exact sequences, pull-back and push-out constructions and the three-lemma; some knowledge of functors and natural transformations and the functor *Ext*. Given spaces Y and Z, an *extension of Y by Z* is a Banach space X such that Y is a closed subspace of X and X/Y is isomorphic to Z. For a given couple (Y, Z) of spaces, *Ext*(Y, Z) denotes the set of all extensions of Y by Z modulo a suitable equivalence relation. If **Ban** denotes the category of Banach spaces and **Set** that of sets, the correspondence is done in such a way that *Ext*: **Ban**×**Ban** → **Set** is a functor in each component. Category theory provides two methods to obtain the elements of *Ext*(Y, Z): one using projective elements ($l_1(I)$-spaces) and the other using injective elements ($l_\infty(I)$-spaces). We shall describe the two methods in detail translating the elements of the general theory to the Banach space setting.

For quasi-Banach spaces (in general, not locally convex spaces), categories do not provide with effective ways to calculate *Ext*(Y, Z) since projective or injective spaces do not necessarily exist. A new theory was developed by Kalton [203] and Kalton and Peck [213] to work with quasi-Banach spaces: it is based on the so-called quasi-linear maps, and consists in proving that extensions of two quasi-Banach spaces correspond to quasi-linear maps. We shall describe this approach in correspondence, when the spaces are locally convex, with the categorical point of view .

Chronologically, the seminal breakthrough was Enflo-Lindenstrauss-Pisier negative solution to Palais problem: *if a Banach space X admits a closed subspace isomorphic to a Hilbert space and such that the corresponding quotient is also isomorphic to a Hilbert space, must X be isomorphic to a Hilbert space?* (or, *is to be isomorphic to a Hilbert space a 3SP property?*). Their construction, although related to finite-dimensional properties of Hilbert spaces, relied on the quasi-linear behaviour of some functions. Indeed, two were the ingredients of their counterexample: (1) construction of (in the terminology of this book) 0-linear finite-dimensional maps such that (2) are increasingly far from linear as the dimension increases. Then Ribe [306] and Kalton [203], independent and almost simultaneously, obtained solutions to the $3SP$ problem for Banach spaces, showing that there exists a quasi-Banach space having an uncomplemented copy of $\mathbb{R}$ such that the corresponding quotient space is isomorphic to l_1. Kalton's paper included many other results, such as the proof that an example replacing l_1 by l_p cannot exist. Moreover, an embryo of what shall become the general theory of twisted sums appears there, with a proof that the only way to obtain a nontrivial twisted sum (i.e., not a direct sum) is that the quasi-linear map be at infinite distance of linear maps. Then it came Kalton and Peck' shaking ground paper [213] with a new counterexample to Palais problem and the whole theory about twisted sums, showing that the way in which Enflo, Lindenstrauss and Pisier had done it was essentially the only way (with the freedom to choose the quasi-linear map).

While this chapter contains the general theory, solutions to Palais problem (i.e., concrete ways to obtain nontrivial quasi-linear maps) shall occur in Chapter 3.

1.1   Short exact sequences

For all basic elements of category theory and homological algebra we suggest [169].

The category **Ban** has Banach spaces as objects and continuous linear mappings as arrows. A short exact sequence in **Ban** is a diagram

$$0 \to Y \overset{i}{\to} X \overset{q}{\to} Z \to 0$$

where the image of each arrow coincides with the kernel of the following one. When no further explanation appears, i denotes the injection and q the quotient map. The open mapping theorem ensures that, given a short exact sequence as before, Y is a subspace of X and Z is the corresponding quotient X/Y. In this context, a $3SP$ problem is: let $0 \to Y \to X \to Z \to 0$ be a short exact sequence in which the spaces Y and Z have a given property P. Must the space X also have P?

In the category **Ban** the *kernel* of a (linear continuous) map $\alpha\colon X \to Y$ is the canonical inclusion $i\colon Ker\,\alpha \to X$ (and the *cokernel* is the canonical quotient map $q\colon Y \to Y/\overline{Im\,\alpha}$). Some basic results about short exact sequences (valid in arbitrary categories where kernels and cokernels exist) follow:

Two short exact sequences $0 \rightarrow Y \rightarrow X \rightarrow Z \rightarrow 0$ and $0 \rightarrow Y \rightarrow U \rightarrow Z \rightarrow 0$ are said to be *equivalent* if some arrow $X \rightarrow U$ exists making commutative the squares of the diagram

$$0 \rightarrow Y \rightarrow X \rightarrow Z \rightarrow 0$$
$$\| \quad \downarrow \quad \|$$
$$0 \rightarrow Y \rightarrow U \rightarrow Z \rightarrow 0$$

This definition makes sense thanks to "the 3-lemma" :

The 3-lemma. *Given a commutative diagram*

$$0 \rightarrow A \rightarrow B \rightarrow C \rightarrow 0$$
$$\alpha\downarrow \quad \beta\downarrow \quad \gamma\downarrow$$
$$0 \rightarrow D \rightarrow E \rightarrow F \rightarrow 0$$

with exact rows, if α and γ are injective then β is injective; also, if α and γ are surjective then β is surjective. Consequently, if α and γ are isomorphisms, then also β is an isomorphism.

Proof. Since there is no place for confusion, we call i the injections in the two short exact sequences and q the quotient maps.

(INJECTIVITY) If $\beta b=0$ with $b \in B$ then $\gamma q b=0$ and $q b=0$. Thus $b=ia$ for some $a \in A$, and since $i\alpha a = \beta i a = \beta b =0$, hence $a=0$ and thus $b=ia=0$.

Observe that the exactness of the sequence $0 \rightarrow D \rightarrow E \rightarrow F \rightarrow 0$ at E or F has not been used. This shall be important later.

(SURJECTIVITY) Let $e \in E$. Since γ is surjective, for some $c \in C$, $\gamma c=qe$. Thus, for some $b \in B$ it must happen $\gamma q b=qe$. Since $q(\beta b - e)=0$, $\beta b - e = id$ for some $d \in D$; and since α is surjective, $\alpha a = d$, which implies $\beta i a = i\alpha a = id = \beta b - e$ and then $e = \beta(b-ia)$. □

Observe that the exactness of the sequence $0 \rightarrow A \rightarrow B \rightarrow C \rightarrow 0$ at A or B has not been used. This shall be important later.

A short exact sequence $0 \rightarrow Y \rightarrow X \rightarrow Z \rightarrow 0$ is said to *split* if it is equivalent to the sequence $0 \rightarrow Y \rightarrow Y \times Z \rightarrow Z \rightarrow 0$. Given an arrow α in a category **Cat**, a *section* for α is a right inverse; i.e., an arrow s in **Cat** such that $\alpha s=1$. A *retraction* for α is a left inverse, i.e., an arrow r in the category such that $r\alpha=1$. When no other comment is made, we understand that sections or retractions are so in **Ban**. If they simply are sections or retractions in **Set** then we shall usually refer to them as "selection maps."

Lemma 1.1.a. *The exact sequence* $0 \to Y \to X \to Z \to 0$ *splits if and only if* i *admits a retraction or* q *admits a section.*

Proof. The exactness of the sequence and the open mapping theorem imply that $i(Y)$ is isomorphic to Y. If s is a section of q, then $x - sq\,x \in Ker\,q = Y$ for every $x \in X$; therefore X admits the decomposition $X = Im\,s + Y$ as follows: $x = sq\,x + (x - sq\,x)$. Since $\{0\} = Im\,s \cap Y$, the sum is direct. Finally, $Im\,s$ is a closed subspace of X and, by the open mapping theorem, it is isomorphic to Z.

Analogously, if r is a retraction of j, then the restriction of q to $Ker\,r$ defines an isomorphism whose bounded inverse is a section of q.   □

1.2   The short exact sequence induced by the pull-back square

Let $A: X \to Z$ and $B: Y \to Z$ two operators. The *pull-back* of $\{A, B\}$ is the space

$$\Xi = \{(x, y): Ax = By\}$$

endowed with the relative product topology, together with the restrictions of the canonical projections onto X and Y. The diagram for the pull-back is:

$$
\begin{array}{ccc}
 & \pi_Y & \\
\Xi & \to & Y \\
\pi_X \downarrow & & \downarrow B \\
X & \to & Z \\
 & A &
\end{array}
$$

The pull-back has the universal property that given any space U with maps u_X and u_Y making commutative the diagram

$$
\begin{array}{ccc}
 & u_Y & \\
U & \to & Y \\
u_X \downarrow & & \downarrow B \\
X & \to & Z \\
 & A &
\end{array}
$$

there is a unique arrow $\pi: U \to \Xi$ in such a way that $\pi_X \pi = u_X$ and $\pi_Y \pi = u_Y$. The pull-back is unique up to isomorphisms. Perhaps the simplest example of pull-back is the kernel of a linear operator $T\colon X \to Z$ (i.e. the pull-back of T and $0\colon 0 \to Y$); its categorical definition is: the kernel of an arrow T is an arrow i such that $Ti=0$ (definition) and such that if j is another arrow such that $Tj=0$ then j factorizes through i (universal property).

It is clear that when the map $A-B\colon X \times Y \to Z$ defined by $(A-B)(x, y)=Ax-By$ is onto, the "diagonal"

$$0 \to \varXi \to X \times Y \overset{A-B}{\to} Z \to 0$$

of the pull-back square

$$
\begin{array}{ccc}
 & \overset{\pi_Y}{\to} & Y \\
\varXi & \searrow & \\
\pi_X \downarrow & X \times Y & \downarrow B \\
 & \searrow & \\
X & \underset{A}{\to} & Z
\end{array}
$$

is a short exact sequence.

This method for obtaining a *special* short sequence is very useful to construct counterexamples to certain 3*SP* problems. Two typical scenarios could be the counterexample to the Dunford-Pettis property (see 6.6.f) and Dierolf's constructions [86]. Dierolf credits this construction to Pisier. This assignation, however, is not exact: the Kislyakov-Pisier method (see [221]) is based upon the push-out construction, which we describe in the next paragraph, and not on the pull-back. In fact, it seems that it was Dierolf who first used a pull-back construction in the topological vector space setting (see [85, 28, 84, 86] and Ostrovskii [272] in the Banach space setting.

1.3 The short exact sequence induced by the push-out square

The dual notion (in the categorical sense) of the pull-back is that of *push-out*. Let $A\colon Y \to X$ and $B\colon Y \to K$ be two operators. Let Δ be the closure of the set

$$D=\{(Bm, Am)\colon m \in Y\} \subset K \times X$$

The push-out Σ of $\{A, B\}$ is the quotient space

$$\Sigma = (K \times X)/\Delta$$

together with the canonical maps $j_X: X \to \Sigma$ given by $x \to (0, x)+\Delta$; and $j_K: K \to \Sigma$ given by $k \to (k, 0)+\Delta$. We shall apply the push-out construction to couples $\{i, T\}$ where T is a continuous operator and i an isomorphism into, thus D itself is closed. Observe that when $i: X \to X$ is an isomorphism into and $T: X \to X$ then Δ is isomorphic to X. The diagram for the push-out is:

$$
\begin{array}{ccc}
 & A & \\
Y & \to & X \\
B\downarrow & & \downarrow j_X \\
K & \to & \Sigma \\
 & j_K &
\end{array}
$$

The push-out has the universal property with respect to this square: given any space U with maps u_X and u_K making commutative the diagram

$$
\begin{array}{ccc}
 & A & \\
Y & \to & X \\
B\downarrow & & \downarrow u_X \\
K & \to & U \\
 & u_K &
\end{array}
$$

there is a unique arrow $j: \Sigma \to U$ in such a way that $jj_X = u_X$ and $jj_K = u_K$. The push-out is unique up to isomorphisms. Perhaps the simplest example of push-out is the cokernel of a linear operator $T: X \to Z$ (i.e. the push-out of T and $0: X \to 0$) that is the quotient map $Z \to Z/\overline{TX}$. The sequence $0 \to \Delta \to K \times X \to \Sigma \to 0$ is obviously exact. We are more interested in another exact sequence.

Proposition 1.3.a. *Let $0 \to Y \to X \to Z \to 0$ be a short exact sequence and let $T: Y \to K$ be an operator. If Σ denotes the push-out of the diagram with arrows i and T, then the sequence $0 \to K \to \Sigma \to Z \to 0$ is exact.*

Proof. The canonical arrow $K \to \Sigma$ given by $k \to (k, 0)+\Delta$ is injective since $(k, 0) \in \Delta$ implies $k=0$; and it is continuous since $\|(k, 0)+\Delta\| \le \|k\|$. The arrow $\Sigma \to Z$ is well defined by $(k, x)+\Delta \to qx$ since the image of elements of Δ is 0 because of $qi=0$. It is surjective since q is surjective and its kernel is

$$\{(k, x)+\Delta: x \in \text{Ker } q, \ k \in K\} = \{(k-Ty, 0)+K : y \in Y, k \in K\}$$

$$= \{(k, 0) + \Delta : k \in K\}. \qquad \Box$$

The approach of Kislyakov. The way Kislyakov [221] reads the push-out construction is: *Given a subspace Y of a Banach space X, and an operator u: Y → M there is a Banach space Σ and an extension U:X → Σ of u, in such a way that M is a subspace of Σ and Σ/M = X/Y.* In other words, he sees the push-out as an extension result: Σ extends M in the same way as X extends Y.

This extension has some regularization properties:

Lemma 1.3.b. *Let* $i: Y \to X$ *be an isometry into and let* $u: Y \to M$ *be an into isomorphism. If* Σ *is the push-out of* $\{i, u\}$ *and* $U: X \to \Sigma$ *is the extension of u then* $\| U \| \leq 1$ *and* $\| U^{-1} \| \leq max \{1, \| u^{-1} \| \}$.

Proof. Firstly observe that U must be an isomorphism by the three-lemma. The assertion about the norm of U is simple:

$$\| Ux \| = \| (0, x) + \Delta \| = inf \{ \| (uy, x+y) \| : y \in Y\} \leq \| x \|$$

with the obvious choice $y=0$. To prove the assertion about U^{-1}, notice that $(m, x) + \Delta = (0, x - u^{-1}m) + \Delta$, and thus all the elements of Σ can be written as $(0, x) + \Delta$. Thus,

$$U^{-1}((m, x) + \Delta) = U^{-1}((0, x - u^{-1}m) + \Delta) = x - u^{-1}m.$$

If $\| (0, x) + \Delta \| < 1$ then there is some $y \in Y$ such that $\| uy \| + \| x+y \| < 1$, therefore

$$\| x \| \leq \| x+y \| + \| y \|$$
$$\leq \| x+y \| + \| u^{-1} \| \, \| uy \|$$
$$< max \{1, \| u^{-1} \| \},$$

and this proves the assertion. $\qquad \Box$

Another interesting fact about push-out squares is:

Proposition 1.3.c. *Let* $Y \to X$ *be an isometric embedding and* $u: Y \to M$ *a continuous operator. If*

$$
\begin{array}{ccc}
Y & \to & X \\
u\downarrow & & \downarrow \\
M & \to & \Sigma
\end{array}
$$

is a push-out square and E is a closed subspace of M then

$$M \quad \rightarrow \quad \Sigma$$
$$\downarrow \qquad \downarrow$$
$$M/E \quad \rightarrow \quad \Sigma/E$$

is a push-out square (the vertical arrows are the natural quotient maps). Moreover,

$$Y \quad \rightarrow \quad X$$
$$\downarrow \qquad \downarrow$$
$$M \quad \rightarrow \quad \Sigma$$
$$\downarrow \qquad \downarrow$$
$$M/E \quad \rightarrow \quad \Sigma/E$$

is a push-out square (diagram).

The proof is just a manifestation of the universal property of the push-out.

1.4 · Extensions of Banach spaces

Given a couple of Banach spaces Y and Z, an *extension* of Y by Z is a Banach space X such that there exists an exact sequence $0 \rightarrow Y \rightarrow X \rightarrow Z \rightarrow 0$. Two extensions of the Banach spaces Y and Z are said to be equivalent if the corresponding short exact sequences are equivalent (see section 1.1.) The (bi)functor assigning to a couple of Banach spaces the set of equivalence classes of extensions shall be denoted *Ext*.

There are two ways to obtain $Ext(Y, Z)$. One is using the projective objects in **Ban** and other using the injective objects (see next paragraph). A Banach space P is said to be *projective* if every exact sequence $0 \rightarrow Y \rightarrow X \rightarrow P \rightarrow 0$ splits. We pass to describe the method using projective spaces.

Lemma 1.4.a. *Every Banach space is linearly isometric to a quotient of some space $l_1(I)$, where I is a set having the same cardinality as a dense subset of X.*

Proof. Let $(x_i)_{i \in I}$ be a dense subset of the unit ball of X. The map $Q: l_1(I) \rightarrow X$ defined by $Q(\lambda) = \Sigma \lambda_i x_i$ is obviously linear and continuous. Since $Q(Ball(l_1(I)))$ is dense in $Ball(X)$, Q is onto, and thus a quotient map. $\square$

The same construction yields the so-called lifting property of $l_1(I)$ spaces: every quotient map $Q: X \rightarrow l_1(I)$ admits a section and thus $l_1(I)$ is actually a complemented subspace of X. Since a projective space must be a complemented subspace of some

$l_1(I)$, and it can be proved that every complemented subspace of $l_1(I)$ is again a $l_1(J)$ space: the only projective Banach spaces are $l_1(I)$-spaces.

Let now $0 \to Y \to X \to Z \to 0$ be an exact sequence, where i and q are respectively, as usual, the names for the injection and quotient map. Let a short exact sequence $0 \to K \to l_1(I) \to Z \to 0$ be called a *projective presentation* of Z. In what follows we shall write l_1 instead of $l_1(I)$ for the sake of simplicity. The quotient map $l_1 \to Z$ shall be denoted p, and the injection $Ker\ p \to l_1$ by j. By the lifting property of l_1, the map p can be lifted to a map $P: l_1 \to X$ with $qP=p$. Since $qPj = pj = 0$, by the universal property of the kernel Y, the map Pj factorizes as $Pj = i\phi$ through i. This gives an arrow $\phi: Ker\ p \to Y$ making the diagram

$$
\begin{array}{ccccccc}
0 \to Y & \xrightarrow{i} & X & \xrightarrow{q} & Z & \to 0 \\
\phi \uparrow & & P \uparrow & & \| \\
0 \to Ker p & \xrightarrow[j]{} & l_1 & \xrightarrow[p]{} & Z & \to 0
\end{array}
$$

commutative.

Conversely, given a map $\phi: Ker\ p \to Y$ it is possible to construct an extension of Y by Z as the push-out of the maps (ϕ, j), i.e., the quotient space

$$T(\phi, j) = (Y \times l_1)/\Delta,$$

where Δ denotes the subspace $\Delta = \{(\phi k, jk): k \in Ker\ p\}$. If $\phi=0$ then obviously $T(\phi, j) = Y \times Z$.

Proposition 1.4.b. *The map ϕ can be extended to a map $\Phi: l_1 \to Y$ if and only if the short exact sequence $0 \to Y \to T(\phi, j) \to Z \to 0$ splits.*

Proof. Let us denote $T = T(\phi, j)$. As usual, $i: Y \to T$ is the canonical injection and $q: T(\phi, j) \to Z$ the quotient map. It is obvious that if the sequence

$$0 \to Y \to T(\phi, j) \to Z \to 0$$

splits, then ϕ has an extension to l_1 given by the composition of the retraction r of i with the lifting P of p to l_1 as in the following diagram:

$$
\begin{array}{ccccccc}
& & \xrightarrow{r} & & \xrightarrow{q} \\
0 \to Y & \leftrightarrows & T(\phi,j) & & Z & \to 0 \\
\phi \uparrow & & P \uparrow & & \| \\
0 \to Ker p & \xrightarrow[j]{} & l_1 & \xrightarrow[p]{} & Z & \to 0
\end{array}
$$

If, conversely, ϕ has an extension Φ to l_1 then $P-i\Phi$ vanishes on $Ker\ p$

$$(P-i\Phi)j = Pj-i\phi = 0;$$

thus, $P-i\Phi$ is well defined on Z. Now the hypothesis enters: when Φ is Y-valued, the map $P-i\Phi$ acting on Z as $\nu(x+Ker\ p) = Px-i\Phi x$ is a section of p. Clearly one has

$$q\nu p = q(P-i\Phi) = qP = p.$$

Therefore $q\nu = id_Z$ since p is surjective. $\square$

Consequently, two maps ϕ, ψ from $Ker\ p$ into Y are said to be *equivalent* if $\phi-\psi$ can be extended to a map $l_1 \to Y$. In this way

Proposition 1.4.c. *There is a correspondence between the sets*

$$\{classes\ of\ equivalence\ of\ extensions\ of\ (Y,\ Z)\}$$

and

$$\{classes\ of\ equivalence\ of\ maps\ \phi\colon K \to Y\}.$$

Trivial extensions are represented by the map $0\colon K \to Y$ that can be easily extended to $0\colon l_1 \to Y$. Nontrivial extensions correspond to maps that cannot be linearly extended to l_1.

Application. One possible application of this categorical tool appears in combination with a result of Johnson and Zippin [197], which prove that continuous maps $T\colon M \to C(S)$ defined on weak*-closed (with respect to duality with c_0) subspaces M of l_1 can be extended to continuous maps $l_1 \to C(S)$. Therefore

Proposition 1.4.d. *Let V be a closed subspace of c_0. Every exact sequence*

$$0 \to C(K) \to X \to V^* \to 0$$

splits.

Proof. The exact sequence

$$0 \to V \to c_0 \to Z \to 0$$

has a dual sequence $0 \to Z^* \to l_1 \to V^* \to 0$ with Z^* weak* closed. Therefore, every linear continuous map $Z^* \to C(K)$ extends to the whole l_1 and by 1.4.b any sequence $0 \to C(K) \to X \to V^* \to 0$ splits. $\square$

It is also possible to construct all extensions of two given Banach spaces using injective spaces. Recall that a Banach space I is said to be *injective* if every exact sequence $0 \to I \to X \to Z \to 0$ splits.

Lemma 1.4.e. *Every Banach space is linearly isometric to some subspace of a* $l_\infty(I)$ *space, where I is a set having the same cardinality as a dense subset of X.*

Proof. Let $(x_i)_{i \in I}$ be a dense subset of X and let $(f_i)_{i \in I}$ be a family of norm one functionals in X^* such that $f_i(x_i) = \| x_i \|$. The map $j: X \to l_\infty(I)$ given by $j(x) = (f_i(x))_{i \in I}$ is obviously a linear isometry into. $\quad\square$

Thus, an injective space must be a complemented subspace of some $l_\infty(I)$ space, (although there exist other injective spaces: Rosenthal constructed in [310] an injective $C(K)$-space that is not a dual space).

The construction via pull-back is probably simpler than that using projective spaces. Let Y and Z be two Banach spaces. Write Y as a subspace of some $l_\infty(I)$ space. We put l_∞ for simplicity. So, we have an exact sequence

$$0 \to Y \to l_\infty \to l_\infty/Y \to 0$$

where j denotes the injection and p the quotient map. Given an operator $T: Z \to l_\infty/Y$ one can obtain an extension of (Y, Z) as follows: if Ξ denotes the pull back of (p, T) then looking at the diagram

$$
\begin{array}{ccccccccc}
& & & & \pi_Z & & & \\
0 & \to & Y & \to & \Xi & \to & Z & \to & 0 \\
& & \| & & \downarrow \pi_\infty & & \downarrow T & & \\
0 & \to & Y & \to & l_\infty & \to & l_\infty/Y & \to & 0 \\
& & & & j & & p & &
\end{array}
$$

it is clear that the sequence $0 \to Y \to \Xi \to Z \to 0$ is exact and thus Ξ is an extension of (Y, Z). Conversely, if $0 \to Y \to X \to Z \to 0$ is an exact sequence and Y is written as a subspace of l_∞, by the extension property, the inclusion $Y \to l_\infty$ can be extended to a map $e_\infty: X \to l_\infty$. Again, looking at the diagram

$$
\begin{array}{ccccccccc}
& & & i & & q & & \\
0 & \to & Y & \to & X & \to & Z & \to & 0 \\
& & \| & & e_\infty \downarrow & & & & \\
0 & \to & Y & \to & l_\infty & \to & l_\infty/Y & \to & 0 \\
& & & & j & & p & &
\end{array}
$$

and since $pe_\infty i = pj = 0$, the map pe_∞ factorizes through q, inducing in this way a map $Z \to l_\infty/Y$. $\quad\square$

Proposition 1.4.f. *There is a correspondence between the sets* {*classes of equivalence of extensions of* (Y, Z) }

and

{*classes of equivalence of maps* $T: Z \to l_\infty(I)/Y$ }

As before, the equivalence relation is: T is equivalent to S if the corresponding extensions are equivalent, and this happens if and only if $T-S: Z \to l_\infty/Y$ can be lifted to a map $Z \to l_\infty$. Trivial extensions are represented by the map $0: Z \to l_\infty/Z$ that can be easily lifted to $0: Z \to l_\infty$. Nontrivial extensions correspond to maps that cannot be continuous and linearly lifted to l_∞.

1.5   Twisted sums of Banach spaces

All the results of this paragraph belong to Kalton [203] and Kalton and Peck [213]

Recall that a *quasi-norm* on a vector space X is an application $\rho: X \to \mathbb{R}^+$ satisfying the same properties of a norm except that the triangle inequality has to be replaced by: for some constant $1 \le K < \infty$,

$$\rho(x+y) \le K[\rho(x)+\rho(y)].$$

A quasi-norm induces a translation invariant metric; a *quasi-Banach* space is a quasi-normed complete vector space.

A *twisted sum of* two Banach spaces Y and Z is a quasi-normed space $Y \oplus_F Z$ (soon we shall explain why this notation) containing a subspace isomorphic to Y such that $(Y \oplus_F Z)/Y$ is isomorphic to Z; i.e., is a quasi-Banach space $Y \oplus_F Z$ for which there exists a short exact sequence

$$0 \to Y \to Y \oplus_F Z \to Z \to 0.$$

When Y and Z are Banach spaces *and also* $Y \oplus_F Z$ is a Banach space then one has a representative of an element of $Ext(Y, Z)$. Representatives of 0 (i.e., equivalent to $Y \oplus Z$) are called *trivial* twisted sums; in this case we shall also say that the twisted sum *splits*.

The main problem to translate all the machinery of categories we explained so far to the category of quasi-Banach or topological vector spaces (in general, not locally convex) is that projective or injective objects do not exist: l_1 spaces are not longer projective (see below) and l_∞ spaces are not longer injective. Here, the approach of Kalton and Peck enters.

A map $F: A \to B$ acting between Banach spaces is said to be *quasi-linear* if it satisfies, for some constant K independent of x and y,

i) $F(\lambda y) = \lambda F(y)$

ii) $\|F(x+y) - F(x) - F(y)\| \le K(\|x\| + \|y\|)$,

Quasi-linear maps arise naturally when considering short exact sequences of (quasi) Banach spaces. Given such a short exact sequence $0 \to Y \to X \to Z \to 0$, a quasi-linear map $F: Z \to Y$ can be obtained as follows:

Consider a linear (possibly non-continuous) selection $L: Z \to X$ and a bounded (possibly non-linear, non-continuous) homogeneous selection $B: Z \to X$ (which exists because the quotient map is open, even in the quasi-Banach case). The difference $F = B - L$ takes values in Y since $q(B - L)=0$ and it is quasi-linear: it is homogeneous because B and L are homogeneous; denoting by C the constant of the quasi-norm and assuming $\|Bx\| \le M\|x\|$ one has

$$\|F(x+y) - F(x) - F(y)\| = \|B(x+y) - B(x) - B(y)\|$$
$$\le C\|B(x+y)\| + C^2(\|B(x)\| + \|B(y)\|)$$
$$\le CM\|x+y\| + C^2M(\|x\| + \|y\|)$$
$$\le 2C^2M(\|x\| + \|y\|).$$

If, conversely, a quasi-linear map $F: Z \to Y$ with constant K is given, it is possible to construct a twisted sum, which shall be denoted $Y \oplus_F Z$, endowing the product space $Y \times Z$ with the quasi-norm

$$\|(y, z)\| = \|y - F(z)\| + \|z\|.$$

This expression is homogeneous and satisfies that $\|(y, z)\| = 0$ implies $(y, z) = 0$ since $F(0)=0$. Moreover:

$$\|(y+u, z+v)\| = \|y+u-F(z+v)\| + \|z+v\|$$
$$= \|y+u-F(z)-F(v)+F(z)+F(v)-F(z+v)\| + \|z+v\|$$
$$\le \|y-F(z)\| + \|u-F(v)\| + \|F(z)+F(v)-F(z+v)\|$$
$$+ (\|z\| + \|v\|)$$
$$\le \|(y, z)\| + \|(u,v)\| + K(\|z\| + \|v\|)$$
$$\le (K+1)(\|(y, z)\| + \|(u, v)\|)$$

shows that it defines a quasi-norm on $Y \times Z$.

Lemma 1.5.a. *The subspace $\{(y, 0): y \in Y\}$ of $Y \oplus_F Z$ is isometric to Y and the corresponding quotient $(Y \oplus_F Z)/Y$ is isometric to Z.*

Proof. Obviously $\| (y, 0) \| = \| y \|$ and this is the subspace part of the statement. As for the quotient,

$$\| (y_0, z) + Y \| = inf \{ \| (y_0, z) + (y, 0) \| : y \in Y \}$$
$$= inf \{ \| y - Fz \| + \| z \| : y \in Y \}$$
$$= \| z \|. \qquad \square$$

Moreover,

Lemma 1.5.b. *Twisted sums of Banach spaces are complete*

Proof. It is easy to see that completeness is a $3SP$ property in the domain of metrizable vector spaces: if $0 \to Y \to X \to Z \to 0$ is a short exact sequence and (x_n) is a Cauchy sequence in X, then (qx_n) is a Cauchy sequence in Z and is therefore convergent. There is no loss of generality in assuming that it converges to 0. Picking y_n in Y so that $d(y_n, x_n) \le 1/n$, it is easy to see that (y_n) is a Cauchy sequence in Y and therefore convergent. This forces (x_n) to be convergent. $\square$

The only thing that remains to show is that the sequences $0 \to Y \to X \to Z \to 0$ and $0 \to Y \to Y \oplus_F Z \to Z \to 0$ are equivalent. If $L: Z \to X$ is a linear (possibly non-continuous) selection for the quotient map, define a linear map $T: X \to Y \oplus_F Z$ by

$$T(x) = (x - Lqx, qx).$$

The map T is continuous since

$$\| Tx \| = \| (x - Lqx, qx) \|$$
$$= \| x - Lqx - Fqx \| + \| qx \|$$
$$= \| x - Lqx - Bqx + Lqx \| + \| qx \|$$
$$= \| x - Bqx \| + \| qx \|$$
$$\le (CM+1) \| x \|$$

and thus, by the 3-lemma, X is isomorphic to $Y \oplus_F Z$.

Therefore, it has been proved that:

Theorem 1.5.c. *There is a correspondence between twisted sums $Y \oplus_F Z$ and quasi-linear maps $F: Z \to Y$.*

The difference between this approach and that using category theory is that extensions of Banach spaces are, by definition, locally convex (Banach spaces) while twisted sums are only quasi-Banach, and thus can be non locally convex.

Moreover, the category of quasi-Banach spaces does not have projective or injective objects and thus the methods developed do not work to obtain quasi-Banach extensions. All this means that nontrivial twisted sums might exist when only trivial extensions exist. An example of this situation is provided by the couple $(\mathbb{R}, l_1)$ for which $Ext(\mathbb{R}, l_1)=0$ while nontrivial twisted sums exist (see 1.6.g).

Thus, it is necessary to develop some criteria to know: a) when twisted sums are locally convex, and b) when a twisted sum is trivial. We start determining when two twisted sums are equivalent following Kalton [203]. We define the distance between two homogeneous maps A and B by

$$dist(A,B) = \sup_{\|x\| \leq 1} \|Ax - Bx\|.$$

We shall denote by $Lin(X, Y)$ the space of all linear (not necessarily continuous) maps between X and Y.

Theorem 1.5.d. *The twisted sums defined by the quasi-linear maps F and G are equivalent if and only if $dist(F-G, Lin(Z, Y)) < +\infty$.*

Proof. Assume that $Y \oplus_F Z$ and $Y \oplus_G Z$ are equivalent. This means the existence of a linear continuous map $T: Y \oplus_F Z \to Y \oplus_G Z$ making commutative the diagram

$$0 \to Y \to Y \oplus_F Z \to Z \to 0$$
$$\| \qquad \downarrow \qquad \|$$
$$0 \to Y \to Y \oplus_G Z \to Z \to 0$$

This commutativity condition yields that $T(y, 0)=(y, 0)$ and $T(0, z)=(Az, z)$ where A is a linear map $Z \to Y$. In this form $T(y, z)=(y+Az, z)$. Therefore:

$$
\begin{aligned}
\|Fz - Gz + Az\|_G &= \|(Fz+Az) - Gz\|_G \\
&\leq \|(Fz+Az, z)\|_G \\
&= \|T(Fz, z)\|_G \\
&\leq \|T\| \, \|(Fz, z)\|_F \\
&= \|T\| \, \|z\|.
\end{aligned}
$$

Conversely, if there exists a linear map $A: Z \to Y$ such that $dist(F-G, A) < \infty$ then $T(y, z)=(y+Az, z)$ is linear and continuous as a map $Y \oplus_F Z \to Y \oplus_G Z$ and makes commutative the diagram above.   □

Since twisted sums split if and only if they come defined by the 0 map, one has.

Corollary 1.5.e. *The twisted sum defined by F is trivial if and only if*

$$dist(F, Lin(Z, Y)) < +\infty.$$

To determine when a twisted sum of Banach spaces is a Banach space, let us introduce three quantities:

$$a_n = \sup_{\|x_i\| \le 1} \|x_1 + \ldots + x_n\|,$$

$$b_n = \sup_{\|x_i\| \le 1} \min_{\pm} \|\sum_{i=1}^{n} \pm x_i\|,$$

$$c_n^2 = \sup_{\sum_{i=1}^{n} \|x_i\|^2 = 1} \frac{1}{n} \frac{1}{2^n} \sum_{\epsilon_i = \pm 1} \|\sum_{i=1}^{n} \epsilon_i x_i\|^2$$

Observe that if

$$Average\ (x_1, \ldots, x_n) = \frac{1}{2^n} \sum_{\epsilon_i = \pm 1} \|\sum_{i=1}^{n} \epsilon_i x_i\|^2,$$

then c_n is the infimum over all constants c such that, for all $x_1, \ldots, x_n$,

$$Average\ (x_1, \ldots, x_n) \le n c^2 \sum_{i=1}^{n} \|x_i\|^2.$$

The relationships of these quantities with the locally convex character of twisted sums appears after three lemmas and a definition:

Lemma 1.5.f. *A quasi-Banach space E is locally convex if and only if*

$$\sup n^{-1} a_n < +\infty,$$

Proof. What has to be proved is that the convex hull of the unit ball is bounded. Let $C > 0$ be a constant for which $a_n \le Cn$. Let $u_1, \ldots, u_n$ be elements in the unit ball, let $\alpha_1, \ldots, \alpha_n$ be positive reals such that $\Sigma_{1 \le i \le n} \alpha_i = 1$, and let m be an integer. If $[m\alpha_i]$ denotes the integer-part of $m\alpha_i$ then:

$$\|\sum_i [m\alpha_i] u_i\| \le a_{\sum_i [m\alpha_i]} \le C \sum_i [m\alpha_i]$$

which implies that

$$\| \sum_i \frac{[m\alpha_i]}{m} u_i \| \leq C.$$

Since $lim_{m \to \infty} [m\alpha_i]/m = \alpha_i$, one has $\| \sum_i \alpha_i u_i \| \leq CK$, where K is the triangle constant of the quasi-norm $\| \ \|$ (it is necessary to do this since quasi-norms are not necessarily continuous with respect to themselves). □

Lemma 1.5.g. ([286]; see also [90, Chapter 13]). *The following conditions are equivalent*:

i) $lim \ n^{-1} b_n \qquad =0$
ii) $sup_{n \geq 2} \ n^{-1} b_n \quad <1$
iii) $lim \ c_n \qquad\qquad =0.$

A quasi-Banach space is said to be *B-convex* if any of those equivalent conditions hold. The equivalence i) ⇔ ii) can be deduced from the submultiplicativity of (b_n) and (c_n) (proved during the next results).

Lemma 1.5.h. *If* $sup \ n^{-1} b_n <1$ *then* $sup \ n^{-1} a_n < \infty$.

Proof. Observe that if (d_n) is a monotone increasing sequence of positive real numbers that is *submultiplicative*, in the sense that $d_{mn} \leq d_m d_n$ and, for some index k, $d_k < k$ then for some $p > 1$, some constant d and large $n \in \mathbb{N}$

$$d_n \leq d \sqrt[p]{n}.$$

To see this, assume that it is $d_2 < 2$. If p is chosen so that $d_2 \leq \sqrt[p]{2}$ then the formula is valid for $n=2^k$. Thus, if $m \in \mathbb{N}$ and k is the greatest integer for which $2^k \leq m$ then

$$d_m \leq d_{2^{k+1}} \leq d_1 \sqrt[p]{2^{k+1}} = d_1 \sqrt[p]{2} \sqrt[p]{2^k} \leq d_1 \sqrt[p]{2} \sqrt[p]{m}$$

and the assertion is proved.

Now, it is easy to prove that the sequences (a_n) and (b_n) are submultiplicative. The proof for (a_n) is immediate, while the proof for (b_n) is as follows (see [90; lemma 13.5]:

If $\{x_{ij}, \ 1 \leq i \leq m, \ 1 \leq j \leq n\}$ is a set of mn vectors for which

$$b_{nm} \leq min_{\epsilon_{ij} = \pm 1} \ \|\sum_{1 \leq i,j \leq n} \epsilon_{ij} x_{ij}\| + \delta$$

then for some signs $\alpha_{ij} = \pm 1$

$$sup_{1 \leq i \leq m} \, \|\sum_{1 \leq j \leq n} \alpha_{ij} x_{ij}\| \leq b_n \, ,$$

and for some signs $\beta_i = \pm 1$

$$\|\sum_{1 \leq i \leq m} \beta_i \sum_{1 \leq j \leq n} \alpha_{ij} x_{ij}\| \leq b_n \, b_m \, .$$

Hence, taking $\gamma_{ij} = \alpha_i \beta_{ij}$ one has

$$b_{nm} \leq \|\sum_{1 \leq i,j \leq n} \gamma_{ij} x_{ij}\| + \delta \leq b_n \, b_m + \delta \, .$$

Passing to the proof of the lemma, observe that one only needs to prove that the sequence $(n^{-1} a_n)$ is bounded for $n = 2^k$; or, if $r > 0$ is chosen, that $(n^{-1} a_n)^r$ is bounded for $n = 2^k$. Since (b_n) is submultiplicative and monotone increasing and $b_2 < 2$, for some $p > 1$

$$sup_n \, n^{-1/p} \, b_n < +\infty \, ,$$

and thus $n^{-1} b_n \leq C n^{-1/p^*}$ for some constant $C > 0$. If $n = 2^k$ and $r > 0$ has been chosen then

$$\sum_k \left(2^{-k} b_{2^k}\right)^r \leq C \sum_k 2^{-kr/p^*} < +\infty \, .$$

The boundedness of $(n^{-1} a_n)^r$ for $n = 2^k$ now follows from the inequality

$$2^{-(n+1)r} a_{2^{n+1}}^r - 2^{-nr} a_{2^n}^r \leq 2^{-(n+1)r} b_{2^{n+1}} \, .$$

Proof of the inequality; and choice of r: It is well-known that a quasi-norm on X is equivalent to some r-norm (i.e., a quasi-norm satisfying $\|x+y\|^r \leq \|x\|^r + \|y\|^r$), see [225], and it is for this new quasi-norm for which one proves the last inequality. Take $\{x_1, .., x_{2n}\}$ a set of $2n$-vectors with $\|x_i\| \leq 1$ and some choice of signs $\varepsilon_i = \pm 1$ for which

$$\|\sum_{1 \leq i \leq 2n} \varepsilon_i x_i\| \leq b_{2n} \, .$$

Let $A = \{i : \varepsilon_i = +1\}$ and $B = \{i : \varepsilon_i = -1\}$. There is no loss of generality assuming that *card* $A \leq n$. Then

$$\sum_{1 \leq i \leq 2n} x_i = 2 \sum_{i \in A} x_i - \sum_{1 \leq i \leq 2n} \varepsilon_i x_i \, ,$$

hence

$$\left\| \sum_{1 \le i \le 2n} x_i \right\|^r \le 2^r \left\| \sum_{i \in A} x_i \right\|^r + b_{2n}^r \le 2^r a_n^r + b_{2n}^r .$$

Therefore

$$a_{2n}^r \le 2^r a_n^r + b_{2n}^r ,$$

and the desired inequality follows. $\qquad\square$

From the lemma it follows that a quasi-normed B-convex space is isomorphic to a B-convex normed space. The quantities c_n appear now.

Theorem 1.5.i.[203] *A twisted sum of B-convex Banach spaces is a B-convex Banach space.*

Proof. We prove that if $\lim c_n(Y)=0$ and $\lim c_n(X/Y)=0$ then $\lim \sup c_n(X)<1$. This is enough since $b_n \le nc_n$.

Let $\{x_{ij}, 1 \le i \le m, 1 \le j \le n\}$ be a set of mn vectors, let θ_{ij} $(1 \le i \le m, 1 \le j \le n)$ be the first mn Rademacher functions on $[0,1]$ and let r_i, $1 \le i \le m$, be the first m Rademacher functions on $[0,1]$. Observe that

$$Average\ (x_1,\ldots,x_n) = \frac{1}{2^n} \sum_{\epsilon_i = \pm 1} \left\| \sum_{i=1}^{n} \epsilon_i x_i \right\|^2 = \int_0^1 \left\| \sum_{i=1}^{n} r_i(t) x_i \right\|^2 dt.$$

For $1 \le i \le m$, let $u_i(t) = \sum_{j=1}^{n} \theta_{ij}(t) x_{ij}$, and for $0 \le t \le 1$ let

$$A(t)^2 = \int_0^1 \left\| \sum_{i=1}^{m} r_i(s) u_i(t) \right\|^2 ds.$$

One has

$$\int_0^1 \left\| \sum_{i=1}^{m} \sum_{j=1}^{n} \theta_{ij}(t) x_{ij} \right\|^2 dt = \int_0^1 \int_0^1 \left\| \sum_{i=1}^{m} \sum_{j=1}^{n} r_i(s) \theta_{ij} x_{ij} \right\|^2 ds\,dt$$

$$= \int_0^1 \int_0^1 \left\| \sum_{i=1}^{m} r_i(s) u_i(t) \right\|^2 ds\,dt$$

$$= \int_0^1 A(t)^2 dt.$$

Since u_i is a simple X-valued function on $[0, 1]$ it is possible to choose a simple Y-valued function v_i such that, being $q: X \to X/Y$ the quotient map, verifies

$$\| u_i(t) + v_i(t) \| \le 2 \| qu_i(t) \| .$$

Clearly $\| v_i(t) \| \le 3 \| u_i(t) \|$. One has:

$$A(t) \le \left(\int_0^1 \| \sum_{i=1}^m r_i(s) (u_i(t) + v_i(t)) \|^2 ds \right)^{1/2} + \left(\int_0^1 \| \sum_{i=1}^m r_i(s) v_i(t) \|^2 \right)^{1/2}$$

$$\le \sqrt{m} \left[c_m(X) \left(\sum_{i=1}^m \| u_i(t) + v_i(t) \|^2 \right)^{1/2} + c_m(Y) \left(\sum_{i=1}^m \| v_i(t) \|^2 \right)^{1/2} \right]$$

$$\le \sqrt{m} \left[2c_m(X) \left(\sum_{i=1}^m \| qu_i(t) \|^2 \right)^{1/2} + 3c_m(Y) \left(\sum_{i=1}^m \| u_i(t) \|^2 \right)^{1/2} \right];$$

therefore

$$\left(\int_0^1 A(t)^2 dt \right)^{1/2} \le$$

$$\le \sqrt{m} \left[2c_m(X) \left(\sum_{i=1}^m \int_0^1 \| qu_i(t) \|^2 dt \right)^{1/2} + 3c_m(Y) \left(\sum_{i=1}^m \int_0^1 \| u_i(t) \|^2 dt \right)^{1/2} \right]$$

$$\le \sqrt{mn} \left[2c_m(X)c_n(X/Y) \left(\sum_{i=1}^m \sum_{j=1}^n \| qx_{ij} \|^2 \right)^{1/2} + \right.$$

$$\left. + 3c_m(Y)c_n(X) \left(\sum_{i=1}^m \sum_{j=1}^n \| x_{ij} \|^2 \right)^{1/2} \right].$$

Thus

$$c_{mn}(X) \le (2c_m(X)c_n(X/Y) + 3c_m(Y)) \, c_n(X).$$

Since

$$lim_{m \to \infty} \; 2c_m(X)c_n(X/Y) + 3c_m(Y) = 0,$$

then $c_m(X) < 1$ for m large enough. $\square$

1.6   Snarked sums of Banach spaces

As we have already seen, twisted sums are locally convex when the summands are
B-convex. In this section we study when a twisted sum is locally convex in terms
of the quasi-linear map that defines the sum.

In a sense, the locally convex character depends on the behaviour of the function

$$\Delta(x_1, \dots, x_n) = \| F (\sum_{i=1}^n x_i) - \sum_{i=1}^n F(x_i) \|,$$

which measures how non-linearly F acts. If F is a quasi-linear map with constant K acting on a Banach space then the estimate

$$\Delta(x_1,\dots,x_n) \le K\left(\sum_{i=1}^{n} i\,\|x_i\|\right)$$

can be easily obtained by induction: the case $n=2$ is clear, and assuming that the preceding estimate holds for $n-1$ one has

$$\left\|F(\textstyle\sum_1^n x_i)-\sum_1^n F(x_i)\right\| \le \left\|F(\textstyle\sum_1^n x_i)+F(\textstyle\sum_2^n x_i)-F(\textstyle\sum_2^n x_i)-\sum_1^n F(x_i)\right\|$$

$$\le \left\|F(\textstyle\sum_1^n x_i)-F(\textstyle\sum_2^n x_i)-F(x_1)\right\| + \left\|F(\textstyle\sum_2^n x_i)-\sum_2^n F(x_i)\right\|$$

$$\le K(\|\textstyle\sum_2^n x_i\| + \|x_1\|)+K(\textstyle\sum_2^n (i-1)\,\|x_i\|)$$

$$\le K\textstyle\sum_1^n i\,\|x_i\|.$$

Digression: Logarithmically convex spaces. From here, Kalton [204] obtains

$$\Delta(x_1,\dots,x_n) \le C \sum^{n} \|x_i\|\left[1 + \log \frac{\sum^{n} \|x_i\|}{\|x_i\|}\right]$$

and thus a property of twisted sums of quasi-Banach spaces that we mention for the sake of completeness:

Theorem 1.6.a. *A twisted sum of Banach spaces is logarithmically convex, in the sense that there is a constant C such that for any finite set $x_1,\dots,x_n$ one has*

$$\left\|\sum^{n} x_i\right\| \le C \sum^{n} \|x_i\|\left[1 + \log \frac{\sum^{n} \|x_j\|}{\|x_i\|}\right]$$

For the proof just take into account the preceding estimate and

$$\left\|\sum (y_i,z_i)\right\| = \left\|\sum y_i-F(\sum z_i)\right\| + \left\|\sum z_i\right\|$$

$$\le \left\|\sum y_i-\sum F z_i\right\| + \left\|\sum F z_i-F(\sum z_i)\right\| + \left\|\sum z_i\right\|$$

$$\le \sum \|y_i-F z_i\| + \sum \|z_i\| + \Delta(z_1,\dots,z_n)$$

$$= \sum \|(y_i,z_i)\| + \Delta(z_1,\dots,z_n).$$

Returning to our affairs, observe that when F is quasi-linear and $x_1,\dots,x_n$ are points such that $\Sigma x_i = 0$ then the preceding estimate gives

$$\left\| \sum_{i=1}^{n} F(x_i) \right\| \leq K \left(\sum_{i=1}^{n} i \, \|x_i\| \right).$$

Intuitively, local convexity is related to an estimate like

$$\left\| \sum_{i=1}^{n} F(x_i) \right\| \leq K \left(\sum_{i=1}^{n} \|x_i\| \right).$$

Following this approach, let us call 0-*linear* to a homogeneous map $F: Z \to Y$ satisfying that whenever $\Sigma_{1 \leq i \leq n} x_i = 0$ then

$$\left\| \sum_{i=1}^{n} F(x_i) \right\| \leq K \sum_{i=1}^{n} \|x_i\|.$$

It is clear that 0-linear maps are quasi-linear since if x and y are two points and $z = x + y$ then $z - x - y = 0$ and thus

$$\| F(z) - F(x) - F(y) \| \leq K(\|x\| + \|y\| + \|z\|) \leq 2K(\|x\| + \|y\|).$$

It is not true, however, that quasi-linear maps are 0-linear. In fact, we shall prove that 0-linear maps correspond to locally convex twisted sums.

Given Banach spaces Y and Z and a quasi-linear map $F: Z \to Y$ let us define the *Snarked sum* of Y and Z as the product space $Y \times Z$ endowed with the norm induced by the set

$$C = conv(\{ (y, 0) : \|y\| \leq 1 \} \cup \{ (F(z), z) : \|z\| \leq 1 \}).$$

We denote by $Y \oplus_C Z$ this normed space.d

Lemma 1.6.b. *If F is 0-linear with constant K. Then the subspace $Y_0 = \{(y, 0): y \in Y\}$ of $Y \oplus_C Z$ is K-isomorphic to Y and the quotient space $(Y \oplus_C Z)/Y_0$ is isometric to Z.*

Proof of the claim. That Y is K-isomorphic to $Y_0 = \{(y, 0): y \in Y\}$ follows from the definition of C realizing that if $(y, 0) \in C$ then

$$(y, 0) = \Sigma \lambda_i(y_i, 0) + \Sigma \mu_j(F(z_j), z_j)$$

with $\Sigma \lambda_i + \mu_j = 1$, $\lambda_i \geq 0$, $\mu_j \geq 0$, $\|y_i\| \leq 1$ and $\|z_j\| \leq 1$. Therefore

$$\| (y, 0) \| \leq \Sigma \lambda_i + \Sigma \mu_j K \leq K.$$

To prove that $(Y \oplus_C Z)/Y_0$ is isometric to Z, observe that

$$(y, z) + Y_0 = (F(z), z) + Y_0.$$

Thus, if $\|z\| \leq 1$, then $\|(y, z) + Y_0\| \leq 1$. Conversely, if $\|(F(z), z) + Y_0\| < 1$ then for some $y \in Y$, $\|(y + F(z), z)\| < 1$ which yields that $(y + F(z), z) \in C$ and thus z is a convex combination of norm one elements of Z, which makes $\|z\| \leq 1$. $\qquad\square$

Therefore,

Proposition 1.6.c. *If F is 0-linear, the Snarked sum is a twisted sum. In particular, it is a Banach space.*

Moreover,

Proposition 1.6.d. *If a twisted sum $Y \oplus_F Z$ is locally convex then it is a Snarked sum.*

Proof. Since the twisted sum $Y \oplus_F Z$ is locally convex, the unit ball B_F of the quasinorm defined by F contains (a multiple of) and is contained in some convex set D, say $\sigma D \subset B_F \subset D$ for some $\sigma > 0$. Since the points $(y, 0)$ with $\|y\| \leq 1$ and $(F(z), z)$ with $\|z\| \leq 1$ have quasi-norm ≤ 1, they belong to B_F and thus to D. One has $C \subset D$. On the other hand, if $(y, z) \in D$ then $\|y - F(z)\| + \|z\| \leq \sigma^{-1}$ and thus $\|y - F(z)\| \leq \sigma^{-1}$ and $\|z\| \leq \sigma^{-1}$. By writing

$$(y, z) = (y - F(z), 0) + (F(z), z)$$

one concludes $(y, z) \in 2\sigma^{-1} C$. $\qquad\square$

Finally we prove the desired result (see [48]).

Proposition 1.6.e. *A twisted sum of Banach spaces $Y \oplus_F Z$ is locally convex if and only if F is 0-linear.*

Proof. Assume that the twisted sum $Y \oplus_F Z$ is locally convex. It therefore coincides with the Snarked sum $Y \oplus_C Z$ where C is the absolutely convex hull of all points $(y, 0)$ with $\|y\| = 1$ and $(F(z), z)$ with $\|z\| \leq 1$.

Let $z_1, \ldots, z_n$ be elements of Z such that $\Sigma_{1 \leq i \leq n} z_i = 0$. Let us call $d = \Sigma_{1 \leq i \leq n} \|z_i\|$. The points

$$\left(F(d^{-1}z_i), d^{-1}z_i \right) \in d^{-1} \|z_i\| C.$$

Thus

$$\left(\Sigma_{i=1}^n F(d^{-1}z_i), 0 \right) = \Sigma_{i=1}^n \left(F(d^{-1}z_i), d^{-1}z_i \right) \in C$$

Since the subspace $\{(y, 0) : y \in Y\}$ of $Y \oplus_F Z$ is isometric to Y, for some constant $M > 0$ one has

$$\|\textstyle\sum_{i=1}^{n} F(d^{-1}z_i)\| \;=\; \Big\|\Big(\textstyle\sum_{i=1}^{n} F(d^{-1}z_i),0\Big)\Big\|_C \;\leq\; M,$$

that by homogeneity yields

$$\|\textstyle\sum_{i=1}^{n} F(z_i)\| \;\leq\; Md$$

and F is 0-linear.

Conversely, assume that F is 0-linear. If $\|(y, z)\| = \|y - F(z)\| + \|z\| \leq 1$ then both $\|y - F(z)\| \leq 1$ and $\|z\| \leq 1$. By writing $(y, z) = (y - F(z), 0) + (F(z), z)$ one concludes $B_F \subset 2C$. Thus, the quasi-norm induced by F is finer than the norm induced by C, and the formal identity $Y \oplus_F Z \to Y \oplus_C Z$ is continuous. Apply 1.6.c and 1.5.b, together with the 3-lemma, to the diagram

$$
\begin{array}{ccccccccc}
0 & \to & Y & \to & Y \oplus_F Z & \to & Z & \to & 0 \\
  &   & \| &   & \downarrow   &   & \| &   &   \\
0 & \to & Y & \to & Y \oplus_C Z & \to & Z & \to & 0
\end{array}
$$

to get that the sequences $0 \to Y \to Y \oplus_F Z \to Z \to 0$ and $0 \to Y \to Y \oplus_C Z \to Z \to 0$ are equivalent. $\qquad\qquad\qquad\qquad\qquad\qquad\qquad\qquad\qquad\qquad\qquad\qquad\qquad\quad\square$

In combination with corollary 1.5.e we see that if F is a quasi-linear map at finite distance of a linear map then it defines a trivial twisted sum, thus the extension space is locally convex and F is 0-linear. This can be proved directly:

Proposition 1.6.f. *If F is a quasi-linear map such that $d = \mathrm{dist}\,(F, Lin) < +\infty$ then F is 0-linear with constant at most d.*

Proof. Let $x_1,\ldots,x_n$ be elements such that $\Sigma x_i = 0$. From $\|x_i\|\, d \geq \|Fx_i - Lx_i\|$ one deduces

$$\textstyle\sum_i \|x_i\|\, d \geq \sum_i \|Fx_i - Lx_i\| \geq \|\sum Fx_i - \sum Lx_i\| = \|\sum Fx_i\|. \qquad\square$$

Let us show that extensions and twisted sums of Banach spaces can be different. This was almost simultaneously shown by Ribe [306] and Kalton [203]. We shall use Ribe's construction. Precisely,

Theorem 1.6.g. *There exists a nontrivial short exact sequence*

$$0 \to \mathbb{R} \to E \to l_1 \to 0.$$

Proof. The only thing one needs is a quasi-linear not 0-linear map $F: l_1 \to \mathbb{R}$. Let

$$F(x) \;=\; \textstyle\sum_i x_i \log|x_i| \;-\; \Big(\sum_i x_i\Big) \log\Big|\sum_i x_i\Big|$$

defined on the subspace of finite sequences, and assuming $0\log0 = 0$. That F is quasi

-linear (with constant 2) is now checked:

Lemma 1.6.h. *For all real numbers s and t one has*

$$| \, s \log|s| \, + \, t \log|t| \, - \, (s+t) \log|s+t| \, | \; \leq \; |s| + |t|.$$

Proof. If t, $s > 0$ then the left-hand side is:

$$|s \log ((s+t)/s) + t \log ((t+s)/t)| \; =$$

$$= -(s+t)[\, s/(s+t) \log (s/(s+t)) + t/(s+t) \log (t/(t+s)) \,]$$

$$\leq (2/e)(s+t)$$

$$\leq s+t,$$

since $|x \log x| \leq 1/e$ when $0 < x < 1$.

The lemma is therefore proved when s and t have the same sign. When not, assume that $s+t>0$, $t>0$ and $s<0$; replace s by $-s$ and t by $s+t$. Then

$$s\log |s| + t\log |t| - (s+t)\log|s+t| \; \leq \; |s| + |s+t| \; \leq \; |s| + |t|. \qquad \square$$

Now the proof that F is quasi-linear is easy: for each i one has:

$$| \, x_i \log|x_i| \, + \, y_i \log|y_i| \, - \, (x_i+y_i) \log|x_i+y_i| \, | \; \leq \; |x_i| + |y_i|.$$

Summing over i

$$| \, \Sigma_i \, x_i \log|x_i| \, + \, \Sigma_i \, y_i \log|y_i| \, - \, \Sigma_i \, (x_i+y_i) \log|x_i+y_i| \, | \; \leq \; \|x\| + \|y\|.$$

Again by the lemma,

$$| \, \Sigma_i \, x_i \log|\Sigma_i \, x_i| \, + \, \Sigma_i \, y_i \log|\Sigma_i \, y_i| \, - \, \Sigma_i \, (x_i+y_i) \log|\Sigma_i \, x_i+y_i| \, |$$

$$\leq |\Sigma_i \, x_i| + |\Sigma_i \, y_i| \; \leq \; \|x\| + \|y\|$$

All together yields

$$| \, Fx + Fy - F(x+y) \, | \; \leq \; 2 \, (\, \|x\| + \|y\| \,).$$

Now extend f to the whole l_1 by density (see p. 90). The space X is the twisted sum $\mathbb{R} \oplus_F l_1$. To see that the corresponding exact sequence does not split, Ribe directly proofs that $\mathbb{R}$ is uncomplemented (which also proves that the twisted sum E is not locally convex) showing that $(1, 0)$ lies in the convex hull of every neighborhood of 0. We only need to verify that F is not 0-linear: observe that

$$F \left(\Sigma_{i=1}^{n} \, e_i \right) = n \log n$$

while $F(e_n)=0$, $\|e_n\|=1$ and $\|\Sigma^n e_i\|=n$, making impossible the existence of

$K > 0$ such that

$$n \log n = \left\| F\left(\sum_{i=1}^{n} e_i\right) - \sum_{i=1}^{n} F e_i \right\|$$

$$\leq K\left(\left\|\sum_{i=1}^{n} e_i\right\| + \sum_{i=1}^{n} \|e_i\|\right) = 2Kn.$$

All this can be stated as:

Theorem 1.6.i. *To be a Banach space is not a 3SP property.*

It is worth remarking that the existence of a non splitting short exact sequence $0 \to \mathbb{R} \to E \to l_1 \to 0$ also shows that the proof 1.5.i (twisted sums of B-convex Banach spaces are locally convex) cannot be extended to arbitrary Banach spaces. Moreover, this shows that l_1 is not projective in the category of quasi-Banach spaces and the quotient map $id: l_1 \to l_1$ cannot be linearly lifted to E.

As we have already seen, given a quasi-linear map $F: Z \to Y$, the quasinorm $\|(y, z)\| = \|z\| + \|y - Fz\|$ is equivalent to a norm if and only F is 0-linear. The question *when is* $\|(y, z)\| = \|z\| + \|y - Fz\|$ *itself a norm?* can also be answered. Recall from [241] that a map Ω is called *pseudolinear* when it satisfies

$$\|\Omega x + \Omega y - \Omega(x+y)\| \leq \|x\| + \|y\| - \|x+y\|.$$

Yost remarked to us the following result.

Proposition 1.6.j. *The expression*

$$\|(y, z)\| = \|y - \Omega z\| + \|z\|$$

is a norm if and only if Ω is pseudolinear.

Proof. The pseudolinearity of Ω makes this expression a norm without further "renorming." Conversely, if $\|y - \Omega z\| + \|z\|$ is a norm then Ω is homogeneous since, when λ is a scalar,

$$\|\lambda \Omega z - \Omega(\lambda z)\| + \|\lambda z\| = \|(\lambda \Omega z, \lambda z)\| = |\lambda| \, \|(\Omega z, z)\| = |\lambda| \, \|z\|$$

which yields $\lambda \Omega z = \Omega(\lambda z)$.

That Ω is pseudo-linear is very easy to verify:

$$\|\Omega z + \Omega b - \Omega(z+b)\| + \|z+b\| = \|(\Omega z + \Omega b, z+b)\|$$

$$\leq \|(\Omega z, z)\| + \|(\Omega b, b)\|$$

$$= \|z\| + \|b\|. \qquad \square$$

All this gives a complete characterization of the locally convex character of twisted sums in terms of quasi-linear maps [48].

Theorem 1.6.k. *Let Y and Z be Banach spaces and F: $Z \to Y$ a homogenous map. We consider the produci space $Y \oplus Z$ and the map $\| (y, z) \|_F = \| y - Fz \| + \| z \|$.*

   i) $\| \ \|_F$ *is a quasi-norm if and only if F is quasi-linear*
   ii) $\| \ \|_F$ *is equivalent to a norm if and only if F is 0-linear*
   iii) $\| \ \|_F$ *is a norm if and only if F is pseudo-linear.*

1.7   Twisted sums and extensions

Regarding what categorical methods have to say, one has.

Proposition 1.7.a. *If the twisted sum $Y \oplus_F Z$ is locally convex then it is equivalent to the extension defined by the restriction $\phi = Q_{|Ker\ q}$, where $Q: l_1 \to Y \oplus_F Z$ is a lifting of the quotient map $q: l_1 \to Z$.*

Proof. It is enough, by the three lemma, to give a continuous map m making commutative the diagram

$$0 \to Y \to T(\phi, i) \to Z \to 0$$
$$\| \qquad \downarrow m \qquad \|$$
$$0 \to Y \to Y \oplus_F Z \to Z \to 0 .$$

where $T(\phi, i)$ denotes the push-out of ϕ: $Ker\ q \to Y$ and i: $Ker\ q \to l_1$ (see 1.4).

An explicit definition of m can be given as follows: If B is a bounded selection for q then a lifting $Q: l_1 \to Y \oplus_F Z$ of q can be defined by

$$Q(\Sigma \lambda_n e_n) = \Sigma \lambda_n Bqe_n;$$

and m is given by:

$$m(\ (y, l) + \Delta\ ) = (y, 0) + Q(l).$$

Observe that the map m is automatically continuous. The map m also comes defined by the universal property of the push-out: given a space (in this case $Y \oplus_F Z$) and two arrows

$$Q: l_1 \to Y \oplus_F Z \quad and \quad j: Y \to Y \oplus_F Z$$

such that $j\phi = Qi$, there is an arrow

$$m: T(\phi, i) \to Y \oplus_F Z$$

making commutative the diagram

$$
\begin{array}{ccccccccc}
0 & \to & Y & \xrightarrow{j} & Y \oplus_F Z & \xrightarrow{p} & Z & \to & 0 \\
& & \| & & \uparrow m & & \| & & \\
0 & \to & Y & \to & T(\phi, i) & \to & Z & \to & 0 \\
& & \phi \uparrow & & & & \| & & \\
0 & \to & Ker\, q & \underset{i}{\to} & l_1 & \underset{q}{\to} & Z & \to & 0
\end{array}
$$

$\square$

Lemma 1.7.b. *An extension $T(\phi, i)$ induced by a map $\phi: Ker\, q \to Y$ is equivalent to the twisted sum induced by the quasi-linear map $F = \phi F_1$ where $F_1: Z \to Ker\, q$ is a quasilinear map corresponding to $0 \to Ker\, q \to l_1 \to Z \to 0$.*

Proof. Assume that $F_1 = B_1 - L_1$, where B_1 is a homogeneous bounded selection for q and L_1 is a linear selection for q. QB_1 and QL_1 are, respectively, homogeneous bounded and linear selections for the quotient map $p: Y \oplus_F Z \to Z$, and their difference is $QB_1 - QL_1 = Q(B_1 - L_1) = \phi(B_1 - L_1)$. Since twisted sums corresponding to different "versions" of the quasi-linear map that defines a sequence are equivalent, the result is proved. $\square$

By categorical duality it is possible to give a second version of the interplay between twisted sums and extensions. The first result is analogous to 1.7.b and we omit the proof.

Proposition 1.7.c. *The extension induced by the map $T: Z \to l_\infty/Y$ is equivalent to the twisted sum induced by the quasi-linear map $F_\infty T$, where $F_\infty: l_\infty/Y \to Y$ is the quasi-linear map inducing $0 \to Y \to l_\infty \to l_\infty/Y \to 0$.*

The second result follows 1.7.a. The local convexity appears when one uses the extension property of l_∞.

Proposition 1.7.d. *Let $Y \oplus_F Z$ be a locally convex twisted sum of Banach spaces. If $0 \to Y \to l_\infty \to l_\infty/Y \to 0$ is an exact sequence and $e: Y \oplus_F Z \to l_\infty$ is an extension of the inclusion map $Y \to Y \oplus_F Z$ then the twisted sum $Y \oplus_F Z$ coincides with the extension obtained from the map $e_\infty|_Z: Z \to l_\infty/Y$.*

1.8   Applications:
Inductive limits of Banach spaces

The push-out construction has been skillfully exploited by Bourgain and Delbaen [36], Pisier [290, 293] and Pisier and Bourgain [38] to produce some interesting spaces.

The approach of Pisier: inductive limits. Pisier uses the push-out in [290] (in the way of Kislyakov) as one of the basic ingredients for the construction of his counterexample of a cotype 2 Banach space X with the Grothendieck property (all operators from X into a Hilbert space are 1-summing; i.e., $L(X, l_2) = \Pi_1(X, l_2)$) and such that the π and ε topologies coincide in $X \otimes X$. What Pisier needs for his construction is that given a subspace (with some additional properties) S of a cotype q Banach space B, and a cotype q space E, to be able to obtain a cotype q space X containing isometrically E such that all finite-rank operators $u: S \to X$ can be extended to finite rank operators $U: B \to X$ with some uniform bound of norms $\| U \| \leq K \| u \|$.

The basic requirement to start the process is to have a subspace S of a Banach space X in such a way that

(*)   if $q: X \to X/S$ denotes the quotient map, there is a constant $\lambda > 0$
         such that for any finite sequence $z_1, \ldots, z_n$ in X/S there is a sequence
         $x_1, \ldots, x_n$ in X with $qx_j = z_j$ and

$$Average \ \| \Sigma \pm x_j \| \ \leq \lambda \ Average \ \| \Sigma \pm z_j \| .$$

Here the average is taken with respect to all choices of signs. In particular, (*) holds if S is isomorphic to a Hilbert space [290, Prop. 2.1].

The first step is then to prove that when B is of cotype q and S is as above, if $t: S \to E$ is an operator into a cotype q space E then the extension E_1 (push-out of the maps $i: S \to B$ and $T: S \to E$) is of cotype q and the norm of the extension $T: B \to E_1$ of t admits a bound $\| T \| \leq K(S, B, q) \| t \|$, where $K(S, B, q)$ is a constant depending on S, B and the cotype of B.

General construction. Let $(S_i, B_i)_i$ be a family of couples, where S_i is a subspace of B_i as before and all the spaces B_i are of cotype q with cotype constants c_q uniformly bounded: $sup \ c_q(B_i) < +\infty$. Let $t_i: S_i \to E$ be a family of operators into a cotype q space E. Then there is a cotype q space E_1 and extensions $T_i: B_i \to E_1$ in such a way that there is a uniform bound $\| T_i \| \leq K \| t_i \|$ for the estimates of the norms. The idea is to consider the subspace $l_1(S_i)$ of $l_1(B_i)$ and the induced operator $t: l_1(S_i) \to E$ defined by

$$t(x_i) = \Sigma\, t_i(x_i).$$

With this, given a couple (S, B) as before and given a cotype q space E it is possible to obtain a cotype q space E_1 (extension of E and B/S) in such a way that all finite rank operators $f: S \to E$ can be extended to finite rank operators $F: B \to E_1$ with a uniform estimate of the norms $\|F\| \le K \|f\|$. One simply needs to consider as index set the family $\{f_i\}_{i \in I}$ of all finite rank operators $f_i: S \to E$ and apply the preceding result to $S_i = S$, $B_i = B$ and $t_i = f_i$.

Inductive step. To obtain a cotype q space X such that all finite dimensional operators $S \to X$ can be extended to $B \to X$ with a uniform bound on the norm we shall apply the preceding schema inductively. The starting step is just as described. In the next step one should take the family of all finite dimensional operators $S \to E_1$ and repeat the process. Observe that each space E_n is isometric to a subspace of E_{n+1} and that the bound of the norms of the extensions only depends of S, B and the cotype of the spaces, q. Therefore the inductive process does not spoil the estimate. The space X is the inductive limit of the sequence $E \to E_1 \to E_2 \to \ldots$ which can also be described as the completion of $\bigcup E_n$. From this, the proof that all finite rank operators $S \to X$ can be extended to $B \to X$ follows easily by a density argument.

It can be remarked that if E and B are separable then X can be chosen separable. Moreover, one can start not just with one couple (S, B) but with any family (S_j, B_j) satisfying the starting conditions.

The Pisier space. The space X of Pisier is obtained through the preceding construction starting with three couples:

♦ the subspace spanned by the Rademacher functions in $L_1(0, 1)$ (quite well-known to be isomorphic to l_2).

♦ the space H_1 spanned by the functions $\{e^{int} : n \ge 0\}$ in $L_1(T)$ being T the unit circle; although H_1 is not isomorphic to l_2, it can be directly proved [290, Corol. 1.12] that it has the basic property (*).

♦ the subspace of L_1/H_1 spanned by the sequence $exp(-3^n it)_n$; this subspace is known to be isomorphic to l_2; that L_1/H_1 has cotype 2 is a result of Bourgain [34].

A p-summing surjection not p-radonifying. Pisier uses again this method in [293] to construct a Banach space Z for which there exists a surjection $Z \to l_2$ that is p-summing but not p-radonifying, $0 < p < 1$.

Construction of $\mathcal{L}_\infty$-spaces. We describe now how to use this inductive method to obtain, starting with any separable Banach space E, a separable $\mathcal{L}_\infty$-space $\mathcal{L}_\infty(E)$ with some additional properties in such a way that E is isometric to a subspace of $\mathcal{L}_\infty(E)$. The method is due to Bourgain and Delbaen [36] and Bourgain and Pisier [38]. First of all, write $E = \overline{\bigcup E_n}$ as the closure of an increasing union of finite dimensional subspaces.

First step. Consider E_1 as $(1+\varepsilon)$-isomorphic to a subspace S_1 of some $l_\infty^{m_1}$ through some isomorphism $u\colon S_1 \to E_1$. If one wants to arrive to an $\mathcal{L}_{\infty,\lambda}$ space then one should write $1+\varepsilon = \lambda\eta$ and consider that $\|u\| = \eta$ (in this way $\|u^{-1}\| \leq \lambda$). If X_1 is the push-out of the diagram

$$
\begin{array}{ccc}
S_1 & \to & l_\infty^{m_1} \\
u \downarrow & & \downarrow U \\
E_1 & \to & D_1 \\
\downarrow & & \downarrow \\
E & \to & X_1
\end{array}
$$

then one sees that $X_1/E = l_\infty^{m_1}/S_1$ is finite dimensional; moreover X_1 contains a subspace D_1 that is λ-isomorphic to $l_\infty^{m_1}$: indeed, D_1 is the push-out of the upper square and, by the lemma 1.3.b, the extension U of u is an isomorphism with

$$\|U\|\,\|U^{-1}\| \leq max\{1, \|u^{-1}\|\} = \lambda.$$

The inductive step. Once the space D_n has been obtained, consider $\langle D_n \cup E_{n+1}\rangle$ in X_n as $(1+\varepsilon)$ isomorphic to a subspace S_{n+1} of $l_\infty^{m_{n+1}}$ through some isomorphism u_{n+1} with $\|u_{n+1}\| = \eta$. Obtaining the push-out diagram

$$
\begin{array}{ccc}
S_{n+1} & \to & l_\infty^{m_{n+1}} \\
u_{n+1} \downarrow & & \downarrow U_{n+1} \\
\langle D_n \cup E_{n+1}\rangle & \to & D_{n+1} \\
\downarrow & & \downarrow \\
X_n & \to & X_{n+1}
\end{array}
$$

one sees that D_{n+1} is λ-isomorphic to $l_\infty^{m_{n+1}}$ while $X_{n+1}/X_n = l_\infty^{m_{n+1}}/S_n$ is finite dimensional.

The construction. Let X denote the inductive limit of the sequence of spaces X_n, and let V denote the inductive limit of the sequence of spaces D_n. It is clear that V is a $\mathscr{L}_{\infty,\lambda}$-space that contains an isometric copy of E. Following the notation of [38], an isometric embedding $M \to \Sigma$ is said to be η-admissible if there is a push-out square

$$
\begin{array}{ccc}
Y & \to & X \\
u\downarrow & & \downarrow \\
M & \to & \Sigma
\end{array}
$$

where $Y \to X$ is an isometric embedding and $\| u \| \le \eta$. The core of the argumentation is the following beautiful technical result:

Lemma 1.8.a. *Let $0 < \eta < 1$. Let $j_n \colon F_n \to F_{n+1}$ be a sequence of η-admissible isometric embeddings between finite-dimensional Banach spaces. If F denotes the inductive limit of the sequence then F has the Radon-Nikodym property and the Schur property.*

In order to apply this to the situation just described observe that V is not necessarily defined by η-admissible maps, while X (which certainly is defined by η-admissible maps) is not defined by finite-dimensional spaces. Nevertheless, the space that satisfies both conditions at once is V/E: it is defined by the inductive sequence of finite dimensional spaces X_n/E (they are finite dimensional simply because so are X_1/E, X_2/X_1, ...) and the η-admissible maps (apply proposition 1.3.c) $X_n/E \to X_{n+1}/E$. Therefore V/E has the Radon-Nikodym property (*RNP*) and the Schur properties. One has

Proposition 1.8.b. *Given any separable Banach space E, there is a separable $\mathscr{L}_\infty$-space $\mathscr{L}_\infty(E)$ such that $\mathscr{L}_\infty(E)/E$ has the Radon-Nikodym property and the Schur properties.*

Specific choices of E lead to some singular spaces:

- $E = \{0\}$ yields the construction of an $\mathscr{L}_\infty$-space with the *RNP* and the Schur property ([36]).

- $E = L_1$ leads to an $\mathscr{L}_\infty$-space without *RNP* (since L_1 does not have it) and not containing c_0 (since L_1 and $\mathscr{L}_\infty(L_1)/L_1$ do not contain c_0, which is a *3SP* property; see 3.2.e).

- $E = l_2$ allows one to obtain a $\mathscr{L}_\infty$-space that is weakly sequentially complete (since l_2 and Schur spaces are; and *weak sequential completeness* is a *3SP* property; see 4.7.a), with *RNP* (another *3SP* property (see 6.5.b) that l_2 and

Schur spaces have). It can be proved that the projective tensor product of $\mathcal{L}_\infty(l_2)$ with itself contains a copy of c_0 (that the space $\mathcal{L}_\infty(l_2)$ does not contains c_0 is clear since it is weakly sequentially complete).

♦ $E=d(w,1)$ (a Lorentz space) gives a l_1-saturated space (since $d(w,1)$ and Schur spaces are l_1-saturated and this is a $3SP$ property (see method 2.9)) with the Dunford-Pettis property (all $\mathcal{L}_\infty$-space have it) that is not a Schur space (since $d(w,1)$ is not Schur) (see [56] for details).

Appendix 1.9

L-summands

Isometric $3SP$ problems seems, at first sight, to be an empty world. Since equivalent norms in $\mathbb{R}^2$ induce the same starting norm in $\mathbb{R}$, it seems to be a closed road try to make any meaningful question about which properties of Y and Z would, isometrically, pass to an extension X. Nevertheless, some interesting questions can be pointed out. This appendix is devoted to one of the most striking: the construction of semi-L-summands. In chapter 5 we treat renorming and other geometrical problems; and in the Appendix 3.7 about the dual $3SP$ problem the question of *isometric* preduals shall also be considered.

A linear projection $P: X \to X$ on a Banach space X is said to be an *L-projection* if for all $x \in X$,

$$\| x \| = \| Px \| + \| z - Px \| .$$

A subspace Y of a Banach space X is said to be an *L-summand* if it is the range of an *L*-projection, i.e., if for some subspace Z one has $X = Y \oplus_1 Z$, i.e., if there is a unique decomposition $x = y + z$ with, $y \in Y$, $z \in Z$ and $\| x \| = \| y \| + \| z \| .$

Proper semi-*L*-summands

The starting point is a definition that goes back to Lima [240]: A subspace Y of a Banach space X is said to be a *semi-L-summand of X* if every point $x \in X$ admits a unique nearest point $\pi x \in Y$ in such a way that

$$\| x \| = \| \pi x \| + \| x - \pi x \| .$$

A semi-L-summand is *proper* if π is not linear. Nonproper semi-L-summands (i.e., with π linear) are easy to construct. Proper semi-L-summands are slightly harder to find. The simplest example is perhaps $\mathbb{R}1$ (the subspace of constant functions) in the real space $C(K)$ of continuous functions over a compact space K: the map π is defined by

$$\pi(f) = \tfrac{1}{2}(max\ f + min\ f).$$

Recall that a $C(K)$-space is isometric to the space $A(S)$ of continuous affine functions on the weak* compact convex set $S=\{\mu \in C(K)^* : \|\mu\| = \mu(1_K) = 1\}$ of probability measures on K. In general, if S is any compact convex subset of a locally convex Hausdorff space, $\mathbb{R}1$ is a semi-*L*-summand of $A(S)$ with the same expression for π since $A(S)$ is a closed subspace of $C(S)$. Now π is linear on $A(S)$ if and only if S has a center of symmetry. We shall see later that this is more than just an example of a one-dimensional proper semi-*L*-summand.

Other examples of proper semi-*L*-summands can be constructed as the direct product (with the appropriate *L*-norm) of a finite number of semi *L*-summands one of which at least is proper. The first example of a two-dimensional proper semi-*L*-summand that cannot be obtained by this trivial construction was given in [280], and much more general examples were obtained in [241]. It was shown in [279] that a real Banach space can be a proper semi-*L*-summand in its bidual. The existence of proper semi-*L*-summands in complex spaces was a well-known open question for some years and was solved in the affirmative in [21] making a rather complicated construction. Building up on the examples in [279] and [21], it was shown in [308] that also a complex Banach space can be a proper semi-*L*-summand in its bidual. We should remark however that it is an open problem the existence of uncomplemented proper semi-*L*-summands.

In what follows we shall try to explain the relation of semi-*L*-summand to a special kind of quasilinear maps. This relation was the main idea behind [241].

Observe that when every point x in a Banach space X admits a unique nearest point πx in a subspace Y, the mapping π, called the *metric projection* onto Y, satisfies

$$\pi(\lambda x + y) = \lambda \pi x + y$$

for any $x \in X$, $y \in Y$ and any scalar λ since

$$\|\lambda x + y - (\lambda \pi x + y)\| = |\lambda|\|x - \pi x\| = dist(\lambda x + y, Y).$$

It follows that the map $B: X/Y \to X$ defined by $B(x+Y) = x - \pi x$ is a homogenous bounded (recall $\|x - \pi x\| = dist(x, Y) = \|x + Y\|$) selection for the quotient map $q: X \to X/Y$. If L is a linear (possibly non continuous) selection for q then the difference

$$\Omega(x+Y) = x - \pi x - L(x+Y)$$

is a quasilinear (linear if and only if π is linear) map $\Omega: X/Y \to Y$.

Digression. If one prefers, Ω can also be obtained making the composition QB of $B(x+Y)=x-\pi x$ with any linear (not necessarily continuous) projection $Q: X \to Y$, which gives $QB(x+Y)=Q(x-\pi x)=Qx-\pi x$. This form of constructing Ω is not different since $x-Qx$ is a linear selection for the quotient map: on one side,

$$x+y-Q(x+y)=x+y-Qx-y=x-Qx$$

shows that it is well defined, whereas $q(x-Qx)=qx$ shows that it is a selection for q. Thus, $L(x+Y)=x-Qx$ and so $(B-L)(x+Y)=x-\pi x-x+Qx=Qx-\pi x=QB$.

End of the digression.

But the map Ω obtained through a metric L-projection is more than quasi-linear.

Claim: If Y is a semi-L-summand of X then π satisfies

$$\| \pi(a+b)-\pi a-\pi b \| \leq \| a+Y \| + \| b+Y \| - \| a+b+Y \| .$$

Proof. Observe that since π is a semi-L-projection $\| \pi x \| = \| x \| - \| x-\pi x \| $. Thus one has:

$$\| \pi(a+b)-\pi a-\pi b \| = \| \pi (a+b-\pi a-\pi b) \|$$
$$= \| a+b-\pi a-\pi b \| - \| (a+b-\pi a-\pi b)-(\pi(a+b)-\pi a-\pi b) \|$$
$$\leq \| a-\pi a \| + \| b-\pi b \| - \| a+b-\pi(a+b) \|$$
$$= \| a+Y \| + \| b+Y \| - \| a+b+Y \| .$$

and the claim follows. Therefore Ω satisfies

$$\| \Omega(a+Y) + \Omega(b+Y) - \Omega(a+b+Y) \| \leq \| a+Y \| + \| b+Y \| - \| a+b+Y \| .$$

Such a map Ω is called in [241] a *pseudolinear map*. What has been proved is then that if Y is a (proper) semi-L-summand of X then there is a (non-linear) pseudolinear map $\Omega: X/Y \to Y$.

Conversely, if there exists some pseudolinear (nonlinear) map $\Omega: Z \to Y$ then there is a Banach space X such that Y is isometric to a (proper) semi-L-summand of X and X/Y is isometric to Z and Y is a X. The construction is as follows: consider $X = Y \oplus Z$ endowed with the quasi-norm

$$\| (y, z) \| = \| y-\Omega z \| + \| z \| .$$

That Y is isometric to the subspace $\{(y, 0): y \in Y\}$ follows without difficulty. The subspace Y is semi-L-embedded in X and the metric projection is

$$\pi(y, z) = y - \Omega z$$

since $$\| (y, z) - \pi(y, z) \| = \| (\Omega z, z) \| = \| z \|$$

$$\leq \inf \{ \| (y, z) - (y', 0) \| : y' \in Y \}$$

$$= dist((y, z), Y),$$

which also shows that X/Y is isometric to Z.   □

In this situation one says that $X = Y \oplus_\Omega Z$ is a (proper) semi-*L*-sum of Y and Z.

Proposition 1.9.a. *Let Y and Z be Banach spaces. There exists a proper semi-L-sum of Y and Z if and only if there is a pseudolinear non-linear map $\Omega \colon Z \to Y$.*

It is also clear that the existence of a pseudolinear map $\Omega \colon Z \to Y$, Y some Banach space, implies the existence of a pseudolinear map $f \Omega \colon Z \to \mathbb{R}$ by simple composition of Ω with some linear continuous functional $f \in Y^*$. The map $f \circ \Omega$ can be chosen non-linear when Ω is nonlinear: pick points a and b where $\Omega(a+b) - \Omega(a) - \Omega(b) \neq 0$ and then some $f \in Y^*$ such that $f(\Omega(a+b) - \Omega(a) - \Omega(b)) = \| \Omega(a+b) - \Omega(a) - \Omega(b) \|$. Conversely, if some pseudolinear nonlinear map $\Omega \colon Z \to \mathbb{R}$ exists then clearly the same occurs replacing $\mathbb{R}$ by any Banach space Y. This shows:

Proposition 1.9.b. *There is a proper semi-L-sum of Y and Z if and only if there is a proper semi-L-sum of $\mathbb{R}$ and Z.*

Therefore, it is no surprise that everything depends on the structure of the "quotient" space Z. Recall from [362] that the unit ball B of a Banach space is said to be *reducible* if there is a bounded convex closed *nonsymmetric* set C such that $B = C - C$. Lima and Yost also defined *weak*-reducible* sets in a similar fashion. However, they coincide with the reducible sets, as Payá and Rodríguez Palacios proved in [279]; a schema of the proof is as follows: if B is reducible and weak*-closed one can always get $B = C - C$ and C weak*-closed, since if C is not weak*-closed then replacing C by its weak*-closure one still has $B = \overline{C}^{weak *} - \overline{C}^{weak *}$.

Proposition 1.9.c. *There is a nontrivial semi-L-sum of $\mathbb{R}$ and Z if and only if the unit ball of Z^* reducible.*

Proof. Assume that $Ball(Z^*)$ is reducible, i.e., $Ball(Z^*) = S - S$ where S is a weak*-compact nonsymmetric convex subset of Z^*. On one hand, the only place where we know of a pseudolinear map is the metric projection of $C(K)$ onto $\mathbb{R}1$:

$$\pi(f) = \tfrac{1}{2}(max\, f + min\, f).$$

On the other hand, the semi-*L*-sum of $\mathbb{R}$ and Z must be $\mathbb{R} \oplus Z$ endowed with some norm. To play the role of nonsymmetric compact space we have S. Moreover, $\mathbb{R} \oplus Z$ can be seen as the subspace $A(S)$ of $C(S)$. That gives the norm to be put in $\mathbb{R} \oplus Z$:

$$\| (\alpha, z) \| = sup \{ |z^*(z) + \alpha| : z^* \in S\}.$$

Thus, one has the commutative diagram:

$$0 \to \mathbb{R}1 \to \mathbb{R}1 \oplus Z \to Z \to 0$$
$$\| \quad\quad \| \quad\quad \|$$
$$0 \to \mathbb{R}1 \to A(S) \to Z \to 0$$

Since there is a nonlinear (S nonsymmetric) semi-L-projection $\pi: A(S) \to \mathbb{R}1$, the things to verify are that $\mathbb{R}1$ is isometric to the subspace $\{(\alpha,0): \alpha \in \mathbb{R}\}$ of $\mathbb{R} \oplus Z$ (pretty obvious) and that the corresponding quotient space is isometric to Z. Recall that π is the restriction of the metric projection of $C(S) \to \mathbb{R}1$ to $A(S)$ and thus defined by

$$\pi(f) = \tfrac{1}{2}(max\ f(S) + min\ f(S)).$$

Thinking the elements $z \in Z$ as in $C(S)$, one has $\pi(\alpha, z) = \pi(z) + \alpha$, and thus

$$\| (\alpha, z) + \mathbb{R}1 \| = \| (\alpha, z) - \pi(\alpha, z) \| = \| (-\pi(z), z) \|.$$

Let us denote by s_M the point of S where z attains its supremum on S, and by s_m the point of S where it attains its minimum. Observe that $z(s_M) - \pi(z) = \pi(z) - z(s_m)$. Therefore

$$\| (-\pi(z), z) \| = |z(s_M) - \pi(z)|$$
$$= (\tfrac{1}{2})(z(s_M) - z(s_m))$$
$$= (\tfrac{1}{2})sup\ z(S - S)$$
$$= \tfrac{1}{2} \| z \|.$$

One implication is proved.

For the converse implication one has to show that if $<u>$ is a one-dimensional proper semi-L-summand in a Banach space X then the unit ball of $(X/<u>)^*$ is reducible. To this end, assume that $\| u \| = 1$ and consider the weak* compact convex set

$$D = \{x^* \in X^* : \| x^* \| = 1 = x^*(u) = 1 \}.$$

Claim. Ball(X^*) = conv ($D \cup -D$).

Since this convex hull is weak*-closed, the Hahn-Banach theorem tells us that it is enough to show that for all $x \in X$

$$\| x \| = max \{ |x^*(x)| : x^* \in D \}.$$

Consider first the case $\pi x = 0$ and assume without loss of generality that $\| x \| = 1$. Then, given an element $z = \alpha u + \beta x \in <u, x>$ one has

$$\| \alpha u + \beta x \| = \| \pi z \| + \| z - \pi z \| = \| \alpha u \| + \| \beta x \| = |\alpha| + |\beta|,$$

which implies that the functional $f \in <u, x>^*$ defined by $f(\alpha u + \beta x) = \alpha + \beta$ has norm 1. Let $F \in X^*$ an extension of f with the same norm; $F \in D$ and $F(x) = 1$. In the general case, we write $x = \lambda u + v$ with $\pi v = 0$ and assume without loss of generality that $\lambda > 0$. The preceding argument gives some $V \in D$ such that $V(v) = \| v \|$ and so $V(x) = \lambda + \| v \| = \| x \|$ and the claim is proved.

If $z^* \in (X/<u>)^* = <u>^{\perp}$ then $z^* = \theta a^* - (1-\theta) b^*$ with $a^*, b^* \in D$ and $0 \le \theta \le 1$. But since $0 = z^*(u) = \theta a^*(u) - (1-\theta) b^*(u) = 2\theta - 1$, necessarily $\theta = 1/2$ and $z^* \in 1/2 (D - D)$. Then, the set that reduces $Ball(Z)$ is

$$C = (1/2) (D - x^*)$$

for some fixed element $x^* \in D$, since clearly $Ball(<u>^{\perp}) = C - C$; moreover, C is a convex, closed and bounded subset of $<u>^{\perp}$. Since $<u>$ is a proper semi-L-summand, D is not symmetric, and C is nonsymmetric too. Hence the unit ball of $(X/<u>)^*$ is reducible.   □

Putting together 1.9.b and 1.9.c one has.

Proposition 1.9.d. *Given Banach spaces Y and Z there exist a proper semi-L-sum of Y and Z if and only if the unit ball of Z^* is reducible.*

Let us remark that the proof in 1.9.c gives an into isometry $X \to A(S)$ sending Y into the constant functions. By an easy application of the Banach-Dieudonné theorem, one verifies that it is onto. Therefore, the subspace $\mathbb{R}1$ of constant functions in $A(S)$ is (up to isometries) the only example of a one-dimensional semi-L-summand in a real Banach space ([240, Corollary 7.4]).

To continue with the connections with the theory of twisted sums, let us consider again the problem of the existence of uncomplemented proper semi-L-summands. Recall that the twisted sum defined by the quasi-linear map F splits if and only if F is at finite distance of a linear map. Analogously, let $\Omega: Z \to Y$ be a nontrivial pseudolinear map.

Proposition 1.9.e. *The subspace Y is complemented in $Y \oplus_\Omega Z$ if and only if*

$\Omega = T+A$, *where T is linear and A is continuous.*

Proof. We know from the general theory that if Y is complemented in $Y \oplus_\Omega Z$ then there is a linear map T such that $dist(\Omega,T) < +\infty$ or, what is the same, $A = \Omega - T$ is bounded in the unit ball of Z. Let $M = sup\{\, \|Az\| : \|z\| \le 1\}$. Since both Ω and T are homogeneous so is A and, therefore, continuous at 0: $\|Az\| \le M\|z\|$. Since Ω and T are both pseudolinear, so is A. This makes it continuous everywhere:

$$\|A(z+v)-Az\| = \|A(z+v)-Av-Az\| + \|Av\|$$
$$\le \|z\| + \|v\| - \|z+v\| + M\|v\|.$$

The converse is equally simple since a continuous map A is (by continuity at 0) bounded on some ball centered at 0; if it is homogeneous, it is bounded on any ball centered at 0. Thus, if $\Omega = T+A$ then $\Omega-T=A$ is bounded, Ω is at finite distance from a linear map, and the twisted sum it defines splits.   □

Let us remark that it is an open problem if uncomplemented proper semi-L-summands exist. The following partial answer seems to be the best available result:

Proposition 1.9.f. *Assume that a nontrivial sequence $0 \to Y \to X \to Z \to 0$ exists, where Y is a semi-L-summand in X. Then Y is uncomplemented in Y^{**}.*

Probably the nicest proof appeared in [45]: If $\Omega: Z \to Y$ is pseudo-linear then

$$\|\Omega(z+v)-\Omega v\| \le 2\|z\| + \|\Omega v\|.$$

Let $L \in l_\infty(Z)^*$ be a Banach limit; i.e., $L((z)_{z \in Z})=L((z+z_0)_z)$ for every $z_0 \in Z$. Assume that $P: Y^{**} \to Y$ is a continuous projection. The map $F: Z \to Y$ given by

$$F(z_0) = P(\,(L(\,(<y^*, \Omega(z+z_0)-\Omega(z)>)_{z \in Z})_{y^* \in Y^*}\,)$$

is linear and at finite distance from Ω.

Hahn-Banach smooth spaces

A subspace Y of a Banach space X is said to be *Hahn-Banach smooth* (*HBS*) if continuous functionals on Y admit a unique norm preserving extension to X. It is easy to see that being *HBS* in the bidual is hereditary. Also, Theorem 2.1 in [18] asserts that X is *HBS* in X^{**} if and only if all separable subspaces are. The following results are in [18].

Proposition 1.9.g. *If X is non-reflexive and Hahn-Banach smooth in X^{**} then X has* *no nontrivial L-projections.*

Proof. Otherwise, assume that $X = Y \oplus_1 Z$ is a nontrivial L-decomposition. Since X is not reflexive, it can be assumed without loss of generality that Y is nonreflexive. Since $X = Y \oplus_1 Z$ is *HDS* in $X^{**} = Y^{**} \oplus_1 Z^{**}$, it is *HBS* in $Y^{**} \oplus_1 Z$ as well. Let $(y^*, z^*) \in Y^* \times Z^*$ with $0 < \| y^* \| < 1$, $\| z^* \| = 1$. Let $f \in Y^{***}$ be a norm preserving extension of y^* and $0 \neq g \in Y^{\perp}$ with $\| g \| < 1 - \| f \|$. It turns out that $\| (f \pm g, z^*) \| = max \{ \| f \pm g \|, \| z^* \| \} = 1$ and thus $(f \pm g, z^*)$ are two different norm preserving extensions of (y^*, z^*). $\square$

As a corollary, one has that this is a geometric property, not stable by finite products: although Y and Z have it, $Y \oplus_1 Z$ does not.

Theorem 1.9.h. *To be Hahn-Banach smooth in X^{**} is not a 3SP property.*

Proof. If X is *HBS* and non-reflexive, the product $X \oplus_1 X$ admits L-projections and it cannot be *HBS*. $\square$

Appendix 1.10

Harte's problem and Taylor's spectrum

In [163], R. Harte posed the following

Problem. *Let X be a Banach space and let A and B operators in X such that there is an exact sequence*

$$0 \to X \to X \times X \to X \to 0$$

where $i(x)=(Ax, Bx)$ and $q(x, y)=Bx-Ay$. Does this sequence split?

Observe that the hypothesis implies that A and B are commuting operators, i.e., $AB=BA$. It is not difficult to see that the answer is yes when the space X satisfies some extension or lifting property, as in the case of l_1, c_0, l_∞ or l_2. Besides, it is not difficult to find short exact sequences $0 \to X \to X \times X \to X \to 0$ that do not split. The first example is due to Harte [163]:

Let G_n be a sequence of finite dimensional Banach spaces that is dense in the class of all finite dimensional Banach spaces with respect to the Banach-Mazur metric. Take $X=l_2(G_n)$ and select sequences (Y_n) and (X_n) of finite dimensional subspaces of X such that $Y_n \subset X_n$ and the projection constant of Y_n into X_n tends to infinity with n. Then the sequence

$$0 \to l_2(Y_n) \to l_2(X_n) \to l_2(X_n/Y_n) \to 0$$

is exact and does not split. Since

$$X = X \times X = X \times l_2(Y_n) = X \times l_2(X_n) = X \times l_2(X_n/Y_n)$$

the sequence

$$0 \to X \times l_2(Y_n) \to X \times l_2(X_n) \to X \times l_2(X_n/Y_n) \to 0$$

does not split, and this sequence is equivalent to some sequence

$$0 \to X \to X \times X \to X \to 0.$$

Moreover, as it was observed by Yost [365], starting with *any* nontrivial sequence $0 \to Y \to X \to Z \to 0$ where Y, X and Z are isomorphic to their square one can obtain a nontrivial sequence $0 \to B \to B \times B \to B \to 0$, namely

$$0 \to Y \times X \times Z \to Y \times X^3 \times Z \to Y \times X \times Z \to 0.$$

Harte's problem is a particular case ($n=2$) of a question (see [224] concluding remarks) about Taylor's joint spectrum for commuting operators [345]: *Is it true that every Taylor-regular n-tuple is splittingly regular?* This problem is relevant in operator theory. The reason is that one of the most important results in spectral theory is the existence of a functional calculus for analytic functions in a neighborhood of the *Taylor spectrum* (see [345] and [346]). However, the construction of the functional calculus is based on a variation of the Cauchy-Weil integral representation formula and it is not explicit. In [353], Vasilescu obtained an explicit formula for the case of operators on a Hilbert space. Recently, Kordula and Müller [224] obtained a formula for Taylor's functional calculus valid for analytic functions in a neighborhood of the *splitting spectrum*.

By the time of closing this book, Müller [262*] showed a surprisingly simple counterexample to these problems that we describe now:

Lemma 1.10.a. *There exists a Banach space X with subspaces Y and M such that $Y+M=X$ for which $D = \{(x, x): x \in Y \cap M\}$ is uncomplemented in $Y \oplus M$.*

Proof. Let K be an uncomplemented subspace of a Banach space B. If X denotes the push-out of the maps α, β: $K \to B$ given by $\alpha y = y$ and $\beta y = -y$ then we already know that $X = B \times B / \Delta$, where $\Delta = \{(x, -x): x \in K\}$, and that the maps i, j: $B \to X$ given by $ib = (b, 0) + \Delta$ and $jb = (0, b) + \Delta$ are isometric embeddings. So, set $Y = iB$ and $M = jB$, which clearly gives $Y + M = X$. Moreover, $Y \cap M = \{(x, 0) + \Delta: x \in K\}$ $= \{(0, x) + \Delta: x \in K\}$, and this space is isometric to K (restriction of i or j to K). Thus, a projection of $Y \oplus M$ onto $\{((k, 0) + \Delta, (0, k) + \Delta): k \in K\}$ would give a projection from Y onto $\{(k, 0) + \Delta: k \in K\}$ or, what is the same of B onto K. $\square$

Example 1.10.b. *There exist a Banach space W and commuting operators A and B in W such that the sequence*

$$0 \to W \to W \oplus W \to W \to 0,$$

where $i(x) = (Ax, Bx)$ and $q(x, y) = -Bx + Ay$, is exact and does not split.

Proof. Let X, Y and M as in the lemma. For i, j integers set

$$W_{i,j} = \begin{cases} Z, & i,j \geq 1; \\ Y \cap M, & i,j \leq 0; \\ Y, & i \geq 1, j \leq 0; \\ M, & i \leq 0, j \geq 1. \end{cases}$$

Clearly $W_{i,j} \subset W_{i+1,j}$ and $W_{i,j} \subset W_{i,j+1}$. Set $W = l_2(W_{i,j})$ and define the shift operators $A, B \in L(W)$: $A((w_{i,j})) = (w_{i-1,j})$ and $B((w_{i,j})) = (w_{i,j-1})$. That A and B are commuting isometries is clear.

Let us verify that the sequence is exact. The map q is surjective:

$$Im\ q = Im\ A + Im\ B = W$$

since $W_{i-1,j} + W_{i,j-1} = W_{i,j}$. The map i is injective:

$$Ker\ i = Ker A \cap Ker\ B = 0.$$

That $qi = 0$ is obvious, so it only remains to show that $Ker\ q \subset Im\ i$. Let $((w_{i,j}), (z_{i,j})) \in Ker\ q$; since for all i, j one has $w_{i,j-1} = z_{i-1,j}$ then

$$w_{i,j} = z_{i-1,j+1} \in W_{i,j} \cap W_{i-1,j+1} = W_{i-1,j}$$

and

$$z_{i,j} = w_{i+1,j-1} \in W_{i,j} \cap W_{i+1,j-1} = W_{i,j-1}.$$

Set $u = (w_{i+1,j}) = (z_{i,j+1})$. Then $iu = (Au, Bu) = ((w_{i,j}), (z_{i,j}))$.

Let us verify that $Im\ i$ is uncomplemented in $W \oplus W$. Assuming that it is complemented with projection P then a projection Q can be defined as

$$Q((w_{i,j}), (z_{i,j})) = (w_{1,0}, z_{0,1}) \in W_{1,0} \oplus W_{0,1}.$$

That Q is a projection is clear since $Q^2 = Q$; moreover, QP is also a projection since $(QP)^2 = Q(PQP) = Q(QP) = QP$, and its range is

$$Im\ QP = \{(w_{1,0}, z_{0,1}): w_{1,0} = z_{0,1} \in W_{0,0}\} = \{(x, x): x \in Y \cap M\}.$$

This space would therefore be complemented in $W_{1,0} \oplus W_{0,1} = Y \oplus M$, in contradiction with the lemma.   □

It is not difficult to adapt this example to obtain, for every $n \geq 2$ a Taylor-regular n-tuple that is not splittingly regular.

Chapter 2

Methods to obtain 3*SP* ideals

In this chapter we present several methods to obtain 3*SP* properties. Most of them seem to be particular cases of the following: a natural transformation between two functors $\tau\colon R \to L$ with R right exact and L left exact; results on lifting of sequences, which in turn include many instances of 3*SP* problems (see [148]), can be seen as good examples where this schema appears. In addition to that, other methods available are: the incomparability method [6, 7, 146], Jarchow's factorization method [184], the opening or gap and an overall view of 3*SP* properties called *P-by-Q* properties [63].

From another point of view, the ultimate word for any positive 3*SP* result seems to be the existence of some kind of selection for the quotient map. The fact that these selections need not be continuous or linear makes it difficult to represent them as elements in the category **Ban**.

Before passing to the methods, some words about the use of space ideals. Three-space problems have been studied mainly for isomorphic properties, most of which may be shown to be stable under the formation of finite products, complemented subspaces, and enjoyed by finite dimensional spaces. These are the characteristic properties of *space ideals* in the sense of [285]. A space ideal is said to be a three-space ideal (shortened to 3*SP* ideal) if whenever Y and X/Y belong to the space ideal so does X. In this context, a general open problem is how to characterize 3*SP* ideals.

Isometric problems have been considered in the appendix to Chapter 1. Properties not stable by finite products occupy an intermediate position: on one side, it is clear that if Banach spaces *Y* and *Z* having property *P* exist such that *Y*×*Z* does not have *P* then a counterexample to the corresponding 3*SP* problem is at hand; and, nevertheless, it may happen that stability by finite products turns out to be the difficult thing to prove. Three cases of seemingly harmless properties not stable by finite products: *not containing isomorphic copies of L_1*, Talagrand's celebrated example [342], *the surjective Dunford-Pettis* property of Leung [236], proved by Bombal, Cembranos and Mendoza [27] to be non stable by products, and *property (w)* of Saab and Saab [318]. For other properties, such as the *Krein-Milman* property, it is still unknown whether they are stable by products.

2.1   Three-space ideals

Here we give a basic description of the language of operator and space ideals. For a more sound background about operator ideals and procedures, we suggest to consult reference [285].

A *space ideal* is a subclass B of the class **Ban** of all Banach spaces satisfying the following conditions:

i)   Finite dimensional spaces belong to B;
ii)   If *E* and *F* belong to B then *E*×*F* belongs to B;
iii)   Complemented subspaces of a elements of B belong to B;
iv)   The class B is stable under isomorphisms.

An *operator ideal* A is a subclass of the class L of all continuous operators satisfying the following conditions:

i)   Finite dimensional operators belong to A
ii)   A + A ⊂ A;
iii)   L A L ⊂ A

Given an operator ideal A, a space ideal *sp*(A) can be associated by means of:

$$sp(A) = \{ X \in \text{Ban} : id_X \in A \},$$

and given a space ideal B, an operator ideal *op*(B) can be associated to it:

$$op(B) = \{ T \in \mathcal{L} : T \text{ factorizes through some space } X \in B \}.$$

A space ideal B is said to be a 3*SP* ideal if whenever *Y* and *X*/*Y* belong to B then *X* also belongs to B.

2.2   Obtaining new 3*SP* ideals from old ones

An interesting problem, probably too vague to be answered is: give a characterization of 3*SP* ideals or, even worse, of "classes" of spaces satisfying the 3*SP* property. A subordinate question to that: given Λ a 3*SP* class

which procedures $\Lambda \to \Lambda^{new}$ *give 3SP classes?*

We shall present one such procedure that depends of obtaining an exact functor. Later, we shall present a second method (incomparability) that produces 3*SP* ideals starting with almost *any* class.

A *covariant functor* F: $\mathbf{Cat_1} \to \mathbf{Cat_2}$ between two categories is a correspondence assigning to each object $X \in \mathbf{Cat_1}$ an object $FX \in \mathbf{Cat_2}$ and to each arrow $\alpha:X \to Y$ an arrow $F\alpha:FX \to FY$ in such a way that $Fid_X=id_{FX}$ and $F(\alpha\beta) = F\alpha\,F\beta$. A *contravariant functor* F: $\mathbf{Cat_1} \to \mathbf{Cat_2}$ is a correspondence assigning to each object $X \in \mathbf{Cat_1}$ an object $FX \in \mathbf{Cat_2}$ and to each arrow $\alpha:X \to Y$ an arrow $F\alpha:FY \to FX$ in such a way that $Fid_X=id_{FX}$ and $F(\alpha\beta) = F\beta\,F\alpha$.

A functor F: **Ban** $\to$ **Ban** is said to be *exact* if it transforms short exact sequences into short exact sequences. The next theorem has an obvious proof.

Theorem 2.2.a. *Given an exact functor F in the category of Banach spaces and a 3SP ideal Λ, the space ideal*

$$F\Lambda = \{X \in \mathbf{Ban}: FX \in \Lambda\}$$

is a 3SP ideal.

Let us show some exact functors and the 3*SP* ideals they generate.

The functor Hom. Given a fixed Banach space A the *covariant* functor $Hom(A,.)$: **Ban** $\to$ **Ban** assigns to a Banach space X the space $Hom(A, X)$ of linear continuous operators between A and X. If $T: Y \to X$ then

$$Hom(A, T): Hom(A, Y) \to Hom(A, X)$$

comes defined by $Hom(A, T)(S) = T \circ S$. It is easy to verify that

Lema 2.2.b. *If $0 \to Y \to X \to Z \to 0$ is a short exact sequence then the sequence*

$$0 \to Hom(A, Y) \to Hom(A, X) \to Hom(A, Z)$$

is exact.

The map $Hom(A, X) \to Hom(A, Z)$ is not, in general, surjective. The space A is said to be *projective* when the functor $Hom(A, .)$ is exact.

The *contravariant* functor $Hom(., B)$: **Ban** $\to$ **Ban** assigns to a Banach space X the space $Hom(X, B)$ of linear continuous operators between X and B. If $T: Y \to X$ then

$$Hom(T, B): Hom(X, B) \to Hom(Y, B)$$

comes defined by $Hom(T, B)(S) = S \circ T$. It is easy to verify that

Lema 2.2.c. *If $0 \to Y \to X \to Z \to 0$ is a short exact sequence then the sequence*

$$0 \to Hom(Z, B) \to Hom(X, B) \to Hom(Y, B)$$

is exact.

The map $Hom(X, B) \to Hom(Y, B)$ is not, in general, surjective. The space B is said to be *injective* when the functor $Hom(., B)$ is exact. This is the case when $B = l_\infty(I)$ for some set I by the Hahn-Banach theorem.

The choice $B = \mathbb{K}$ makes $Hom(X, B) = X^*$ and one obtains:

Duality functor and dual ideal.
It is a well-known algebraic fact that the contravariant functor $Hom(., \mathbb{K})$ is exact. By the Hahn-Banach theorem, the duality (contravariant) functor given by $X \to X^*$ and $T \to T^*$ is also exact. We include a proof for the sake of completeness:

Lemma. 2.2.d. *If $0 \to Y \to X \to Z \to 0$ is an exact sequence in the category of Banach spaces, then the dual sequence $0 \to Z^* \to X^* \to Y^* \to 0$ is exact.*

Proof. Since q is surjective, q^* is injective; the Hahn-Banach theorem provides the surjectivity of i^*. It is clear that

$$Ker\ i^* = Y^\perp = (Ker\ q)^\perp = (X/\ Ker\ q)^* = Z^* = (Im\ q)^* = Im\ q^*. \qquad \square$$

As a consequence, the functors $D^n: X \to X^n$ sending a Banach space to its nth-dual are exact. Let α be an even ordinal; i.e., an ordinal having the form $\beta + 2n$ where β is a limit ordinal and $n \in \mathbb{N}$. Then define by transfinite induction $X^{\alpha+2} = (X^\alpha)^{**}$; and if β is a limit ordinal then $X^\beta = \overline{\bigcup_{\alpha < \beta} X^\alpha}$ is the completion of the union of all even duals X^α with $\alpha < \beta$. For odd duals $\alpha + 1$, with α even or limit, $X^{\alpha+1} = (X^\alpha)'$.

Lemma 2.2.e. *Let α be an ordinal. The functor D_α sending a Banach space X to its transfinite α-dual with the corresponding natural induced maps is exact.*

Proof. Only the case of a limit ordinal is not evident. Let β be a limit ordinal and let $0 \to Y \to X \to Z \to 0$ be a short exact sequence. Assume that $\alpha < \beta$. Since the maps induced by $D^{\alpha+2}$ extend those induced by D^{α}, there are well defined maps

$$\bigcup_{\alpha<\beta} Y^{\alpha} \to \bigcup_{\alpha<\beta} X^{\alpha} \to \bigcup_{\alpha<\beta} Z^{\alpha} .$$

The first arrow is clearly an into isometry and the second map takes unit ball onto a dense subset of unit ball. Hence, the continuous extension to the completion provides an exact sequence $0 \to Y^{\beta} \to X^{\beta} \to Z^{\beta} \to 0$. $\qquad\Box$

One has

Theorem 2.2.f. *Given Λ a 3SP ideal, $\Lambda^{\alpha} = \{X : D_{\alpha}X \in \Lambda\}$ is a 3SP ideal.*

Nevertheless, very few things are known about the classes Λ^{α}. For instance, if Λ denotes the class of quasi-reflexive spaces then $\Lambda^{\omega^2} = $ *reflexive spaces*. The reader should consult [122] and [283] for further results.

In [184], Remark 5. (b), Jarchow raises the question: *if Λ^{dual} is a 3SP ideal, is Λ a 3SP ideal?* The answer is no and we shall present several examples: the hereditary Dunford-Pettis property (see 6.6.f), the property of having unit ball Cech complete (see 4.14), the Plichko-Valdivia property (see 4.12) and the property of being, simultaneously, Asplund and *WCG* (see 4.19.d).

Ultrapower functor and ultrapower ideal. General references for ultraproduct theory in Banach spaces are [327, 167]. An ultrafilter on some set J can be understood as a method to linearly assign to families $(x_j)_{j \in J}$ of elements of some compact subset of a topological vector space another element of the compact set that is usually called the *limit of (x_j) by the ultrafilter U*, and denoted $lim_U x_j$. It can also be seen as a family U of subsets of J closed under finite intersections and supersets (if $A \in U$ and $A \subset B$ then $B \in U$), and that is maximal with respect to those properties (which means that if A is any subset of J, either A or $J \backslash A$ belong to U). The ultrafilter is *free* when the intersection of its elements is empty; this implies that the limit by the ultrafilter of convergent families (provided J has some additional order structure) coincides with their usual limit in the topology of the compact space. In this form, given a bounded family $(x_j)_{j \in J}$ of elements of a Banach space X and a free ultrafilter on J, the "limit" $lim_U \| x_j \|$ is well defined. If $l_{\infty}(J, X)$ denotes the space of bounded families of X and *Ker U* denotes the space of bounded families for which $lim_U \| x_j \| = 0$ then the quotient space $X_U = l_{\infty}(J,X)/Ker\ U$ is a Banach space endowed with the norm $\| (x_j) \| = lim_U \| x_j \|$. The space X_U is called the *ultrapower of X by the ultrafilter U*. It obviously contains an isometric copy of X generated by the constant families. If $T: Y \to X$ is a continuous operator, there is an induced operator $T_U: Y_U \to X_U$ defined by

$$T_U((x_j) + Ker\ U) = (Tx_j) + Ker\ U.$$

Lemma 2.2.g. *The covariant functor* $X \to X_U$ *is exact* [184, 269].

Proof. Given a short exact sequence $0 \to Y \to X \to Z \to 0$, the exactness of the induced sequence $0 \to Y_U \to X_U \to Z_U \to 0$ follows from the fact that if $[x_j] \in X_U$ is 0 in Z_U that means $lim_U \|\ x_j + Y\ \| = 0$. Denoting $A_n = \{j:\ dist(x_j, Y) < 1/n\} \in U$, it is possible to pick elements $y_{j,\ n} \in Y$ such that $\|\ x_j - y_{j,\ n}\ \| \leq 1/n$ whenever $j \in A_n \backslash A_{n+1}$. The bounded family (y_j) defined by

$$y_j = y_{j,\ n}\ \text{whenever}\ j \in A_n \backslash A_{n+1}\ \text{and}\ y_j = x_j\ \text{otherwise}$$

verifies that $lim_U\ \|\ x_j - y_j\ \| = 0$, which amounts $[x_j] = [y_j] \in Y_U$.   □

Therefore, denoting $\Lambda_U = \{X \in \mathbf{Ban} : X_U \in \Lambda\}$ one has

Theorem 2.2.h. *Given a 3SP space ideal* Λ, *the ultrapower (by U) ideal* Λ_U *is a 3SP ideal. Also, the ultrapower ideal*

$$\Lambda^{up} = \{X \in B: for\ all\ U,\ X_U \in \Lambda\}$$

is also a 3SP ideal.

If Λ is an injective space ideal (i.e., a space ideal stable by taking subspaces), $X \in \Lambda^{up}$ is the super-property corresponding to $X \in \Lambda$ (see [167]). Recall that given a property P the corresponding super property *super-P* is defined as: every Banach space finitely representable in X has P. The method is thus adequate to study super-properties. For instance, once we know that reflexivity is a 3*SP* property, this method yields that super-reflexivity is a 3*SP* property (see also 4.5).

The converse implication does not hold: let *SC* denote the class of Banach spaces admitting an equivalent strictly convex norm; then *SC* fails the 3*SP* property (see 5.17.d) while *super-SC* = $\{X : X$ admits an equivalent norm such that $X_U \in SC\}$ is equivalent to super-reflexivity (see [2, Thm. II.4.5]), which is a 3*SP* property.

Residual classes. The following method to obtain 3*SP* ideals starting from 3*SP* ideals was introduced in [269].

Lemma 2.2.i. *The functor* $co: X \to X^{co} = X^{**}/X$ *is exact.*

Proof. Let $0 \to Y \to X \to X/Y \to 0$ be a short exact sequence. The proof is based in the isomorphisms (see [70, 269]): $Y^{**}/Y = (X + Y^{\perp\perp})/X$ and $(X/Y)^{**}/(X/Y) = X^{**}/(X + Y^{\perp\perp})$ which immediately give the exactness of the sequence

$$0 \to Y^{**}/Y \to X^{**}/X \to (X/Y)^{**}/(X/Y) \to 0. \qquad \Box$$

Therefore.

Theorem 2.2.j. *Given a 3SP ideal Λ, the residual ideal Λ^{co} is a 3SP ideal.*

This would give another proof that reflexivity is a $3SP$ property: since 0 is a $3SP$ class, $0^{co} = \{reflexive\ spaces\}$ is a $3SP$ ideal.

Question. *It is apparently unknown whether "Λ^{co} is a 3SP class" implies "Λ is a 3SP class."* The following partial consideration can be made.

Theorem 2.2.k. *Assume that P is not a 3SP property and that a counterexample $0 \to X \to X^{**} \to X^{**}/X \to 0$ exists. Assume that there exists some space Z such that $Z^{**}/Z = X$. Then P^{co} is not a 3SP property.*

Proof. Just look at the commutative diagram

$$
\begin{array}{ccccccc}
 & 0 & \to & 0 & & \to & 0 \\
 & \downarrow & & \downarrow & & & \downarrow \\
0 \to & Z & \to & Z^{**} & & \to & X & \to 0 \\
 & \downarrow & & \downarrow & & & \downarrow \\
0 \to & Z^{**} & \to & Z^{****} & & \to & X^{**} & \to 0 \\
 & \downarrow & & \downarrow & & & \downarrow \\
0 \to & Z^{**}/Z & \to & Z^{****}/Z^{**} & \to & X^{**}/X & \to 0 \\
 & \downarrow & & \downarrow & & & \downarrow \\
 & 0 & & 0 & & & 0
\end{array}
$$

with exact arrows and columns. $\qquad \Box$

Note that it was shown in [78] that every *weakly compactly generated* Banach space X there exists some Banach space Z such that $X = Z^{**}/Z$.

Vector-valued functions. The exactness properties of the functors *Hom* have been already mentioned. Less well-known is that given a compact space K the covariant functor $C(K, \cdot)$: **Ban** $\to$ **Ban** that assigns to a Banach space X the Banach space $C(K, X) = \{f: K \to X \text{ continuous}\}$ with the *sup* norm, and to a continuous operator $T: X \to Y$ the continuous operator M_T: $C(K, X) \to C(K, Y)$ given by $M_T(f) = fT$ is exact. The proof is based on the existence of a continuous (usually non linear) selection for the quotient map.

Bartle-Graves continuous selection 2.2.1. *Let $q: X \to Z$ be a quotient map. There exists a continuous map $s: Z \to X$ such that $qs = id_Z$.*

We prove a stronger version taken from [367]:

Let $0 \to Y \to X \to Z \to 0$ be a short exact sequence of Banach spaces. Given $\varepsilon > 0$ there exists a continuous homogeneous selection s for the quotient map such that, for all $z \in Z$,

$$\| s(z) \| \leq (1+\varepsilon) \| z \| .$$

Proof. Let $q: X \to Z$ be the quotient map. One needs only to construct a continuous map h from the unit sphere of Z into X in such a way that $qh(z) = z$ and $\| h(z) \| \leq (1+\varepsilon)$. Since, once this has been done, a "homogeneization" can be performed as follows [96]: let h be a continuous selection for the quotient map; if the space is real, set $d(z) = 1/2(h(z) - h(-z))$; this selection satisfies $d(\lambda z) = \lambda d(z)$ for $|\lambda| = 1$. If the space is complex and μ denotes the normalized Haar measure on the group $T = \{\lambda: |\lambda| = 1\}$ then $d(z) = \int_T \lambda^{-1} h(\lambda z) d\mu(\lambda)$. Take then

$$s(z) = \| z \| d(z/ \| z \|).$$

Returning to the proof, the map h shall be obtained as a limit of maps h_n constructed inductively.

Given $\lambda > 0$, we set $Q(z)_\lambda = \{x \in X: qx = z, \| x \| \leq \lambda\}$. Let $z_0 \in Z$, $\| z_0 \| = 1$. Pick some element $x_0 \in Q(z_0)_\lambda$ and set $h_{1, z_0}(z) = x_0$. Let $V(z_0)$ be a neighborhood of z_0 inside the unit sphere of Z such that for every $z \in V(z_0)$ one has

$$dist\ (x_0, Q(z)_\lambda) < 2^{-1}.$$

From the covering of the unit sphere of Z by neighborhoods $V(z)$ it is possible to select a locally finite refinement $(V_\alpha)_\alpha$ [220, Corol. 5.35]. Let $(u_\alpha)_\alpha$ be a corresponding subordinated partition of the unity [220, Problem 5.W]. The map

$$h_1(z) = \sum_\alpha u_\alpha(z) h_{1, z_\alpha}(z)$$

is continuous and bounded and verifies

$$dist\ (h_1(z), Q(z)_\lambda) < 2^{-1}$$

for all z in the unit sphere of Z because $Q(z)_\lambda$ is convex.

Assume that continuous maps $h_1, \ldots, h_n$ have already been constructed satisfying

$$dist\ (h_j(z),\ Q(z)_\lambda) < 2^{-j}$$

and $\| h_{j-1}(z) - h_j(z) \| < 2^{-(j-2)}$.

By the induction hypothesis, the set

$$U_n = \{x \in Q(z)_\lambda : dist\ (x, h_n(z)) < 2^{-n}\}$$

is non-empty and convex for every $z \in Z$. What one needs to assure is that

$$dist\ (h_{n+1}(z),\ U_n) < 2^{-(1+n)}.$$

This can be done in a neighborhood of every norm one element $z \in Z$ by means of a constant map $h_{n+1,z}$. The map h_{n+1} is constructed "pasting together" these functions by means of a partition of the unity as in the case of h_1. The sequence (h_n) uniformly converges to a function h which satisfies the conditions required.

Proposition 2.2.m. *Let K be a compact space. The functor $C(K, \cdot)$ is exact.*

Proof. Let $0 \to Y \to X \to Z \to 0$ be a short exact sequence. The Bartle-Graves continuous selection s for the quotient map provides the exactness of the functor $C(K, \cdot)$ at Z; i.e., that the induced map $M_q: C(K, X) \to C(K, Z)$ is in fact a quotient map since any $f \in C(K, Z)$ can be written as $M_q(s \circ f)$. Its kernel is $C(K,Y)$. $\qquad\qquad\square$

It is clear that other spaces of functions, such as p-integrable functions over a finite measure compact space, can play the role of $C(K)$. Therefore.

Theorem 2.2.n. *If Λ denotes a 3SP space ideal, the classes*

$$C(\Lambda) = \{spaces\ X\ such\ that\ C(K, X) \in \Lambda\ \}$$
$$L_p(\Lambda) = \{spaces\ X\ such\ that\ L_p(\mu, X) \in \Lambda\ \},\ 1 \leq p \leq \infty,$$

are 3SP ideals.

2.3 Obtaining 3SP ideals

We need some variation of the exact functor method to obtain new 3SP properties. An observation of what happens, for instance, with reflexivity makes clear that properties having the form $X = FX$ where F is an exact functor must be 3SP properties. A precise meaning to the symbol "=" is what natural transformations give.

Let F and G be functors acting between two given categories $\mathbf{Cat_1}$ and $\mathbf{Cat_2}$. A natural transformation $\tau: F \to G$ is a correspondence assigning to an object X in $\mathbf{Cat_1}$ an arrow $\tau X: FX \to GX$ in $\mathbf{Cat_2}$ in such a way that if $\alpha: X \to Y$ is an arrow in $\mathbf{Cat_1}$ then the square

$$
\begin{array}{ccc}
 & \tau X & \\
FX & \to & GX \\
F\alpha \downarrow & & \downarrow G\alpha \\
FY & \to & GY \\
 & \tau Y &
\end{array}
$$

is commutative in $\mathbf{Cat_2}$.

Now, if F and G are exact functors and there is a natural transformation $\tau: F \to G$, the 3-lemma yields that $\{X: \tau X$ is an isomorphism$\}$ is a 3*SP* ideal. This is the case of reflexivity when one considers the natural transformation $\tau: id \to D_2$ between the identity and the bidual functor where $\tau X: X \to X^{**}$ denotes the canonical inclusion.

In many situations, however, one does only have a situation partially behaving as described; for instance, properties involving the coincidence of spaces of sequences or operators, such as the Schur property (weakly convergent sequences are convergent) or the Dunford-Pettis property (weakly compact operators are completely continuous). In these situations, the functors considered might not be exact. Those cases can be covered as follows.

A covariant functor L is said to be *left exact* if exact sequences $0 \to Y \to X \to Z \to 0$ are transformed into left exact sequences $0 \to LY \to LX \to LZ$. A covariant functor R is said to be *right exact* if exact sequences $0 \to Y \to X \to Z \to 0$ are transformed into right exact sequences $RY \to RX \to RZ \to 0$. Analogously in the contravariant case.

Proposition 2.3.a. *Let R and L be right and left exact functors, respectively, from the category **Ban** into another suitable category. Assume that there is a natural transformation $\tau: R \to L$. Then the classes*

$$\{X \in \mathbf{Ban} : \tau X: RX \to LX \text{ is injective}\}$$
$$\{X \in \mathbf{Ban} : \tau X: RX \to LX \text{ is surjective}\}$$
$$\{X \in \mathbf{Ban} : \tau X: RX \to LX \text{ is an isomorphism}\}$$

are 3SP ideals.

Proof. The 3-lemma can be proved in the situation:

$$RY \rightarrow RX \rightarrow RZ \rightarrow 0$$

$$\tau_Y\downarrow \quad \tau_X\downarrow \quad \tau_Z\downarrow$$

$$0 \rightarrow LY \rightarrow LX \rightarrow LZ$$

with horizontal arrows exact (the surjectivity of $E \rightarrow F$ or the injectivity of $A \rightarrow B$ were not used in the proof of the original 3-lemma). $\square$

Some applications:

Reflexivity. Recall that D_2: **Ban** $\rightarrow$ **Ban** is the functor assigning to a Banach space X its bidual X^{**} and to an operator T its bitranspose T^{**}. So, reflexivity is defined by the natural transformation τ: $id \rightarrow D_2$, in the sense that a space X is reflexive when τX is surjective.

When there is no natural dependence it is not clear (to us) what happens; for instance

Question. *Is the class of Banach spaces isomorphic to its bidual a 3SP class? The same question can be asked with respect to the other (transfinite or not) duals. That being isomorphic to its dual* is not a 3SP property was observed by F. Cabello [45]: take a nontrivial exact sequence $0 \rightarrow l_p \rightarrow Z_p \rightarrow l_p \rightarrow 0$ as in 3.2.a. Since $l_{p*}=l_{p*}\oplus l_{p*}$ the sequence $0 \rightarrow l_p\oplus l_{p*} \rightarrow Z_p\oplus l_{p*}\oplus l_{p*} \rightarrow l_p\oplus l_{p*} \rightarrow 0$ is exact and while $l_p\oplus l_{p*}$ is isomorphic to its dual, $Z_p\oplus l_{p*}$ is not (Z_p is isomorphic to its dual) and incomparable with l_{p*}).

Baire-1 functions. The functor B_1: **Ban** $\rightarrow$ **Ban** defined by

$$B_1(X) = \{\text{weak*-limits in } X^{**} \text{ of weakly Cauchy sequences of } X\}$$

is not exact, although it is left exact. Since the functor id is exact, the class of Banach spaces such that $X=B_1(X)$ (weak sequential completeness) is a 3SP ideal.

Intermediate classes method. An *intermediate class* $S = \{S(X)\}_{X \in \text{\textbf{Ban}}}$ is a family of normed spaces such that

i) $X \subset S(X) \subset X^{**}$;
ii) For every pair of spaces Y,X and every operator $T: Y \rightarrow X$, one has $T^{**}(S(Y)) \subset S(X)$.

The class S is said to be *injective* if $S(Y)=(T^{**})^{-1}S(X)$ for every injective isomorphism $T: Y \rightarrow X$. It is said to be surjective if for every surjective operator $T: X \rightarrow Z$, one has $S(Z)=T^{**}S(X)$.

Proposition 2.3.b. *If S_1 is surjective, S_2 is injective and $S_1(X) \subset S_2(X)$ for every X then*

$$\{ X \in \mathbf{Ban} : S_1(X) = S_2(X) \}$$

has the 3SP property.

Proof. This result can be seen as an application of Proposition 2.3.a: the definition of intermediate class says that it defines a functor $X \to S(X)$ and $T \to T^{**}$ (restricted to $S(X)$). Let $0 \to Y \to X \to Z \to 0$ be an exact sequence. Being surjective means that S_1 is exact at Z, and injectivity means that the functor S_2 is exact at X (clearly, the hypothesis on S_2 applies to obtain $u \in i^{**}S_2(Y)$). It was already observed after the proof of the three-lemma that when the property considered is $\{X$: the induced map $S_1(X) \to S_2(X)$ is surjective$\}$ one does not need exactness of S_1 at X. That is what one considers here. One always has exactness at Y. □

The following examples of intermediate classes are taken from [147]:

$B_1(X)$ $\qquad$ = weak* limits in X^{**} of weakly Cauchy sequences of X
$DBSC(X)$ $\qquad$ = weak* limits in X^{**} of wuC series in X
$S_1(X)$ $\qquad$ = weak* sequentially continuous elements of X^{**}
$C(X)$ $\qquad$ = $\{F : \lim F(f_n) = 0$ if $(f_{n+1} - f_n)$ wuC in X^* with (f_n) weak* null$\}$
$GS(X)$ $\qquad$ = $\cup \{M^{\perp\perp} : M$ is a separable subspace of $X\}$

The classes B_1 and $DBSC$ are injective (not surjective); the class GS is surjective; the classes S_1 and C are neither injective nor surjective (see [147]). Observe the translation of classical properties.

X is weakly sequentially complete $\qquad\qquad$ $\Leftrightarrow X = B_1(X)$
X contains no copies of c_0 $\qquad\qquad$ $\Leftrightarrow X = DBSC(X)$
X contains no complemented copy of l_1 $\qquad$ $\Leftrightarrow X^* = DBSC(X^*)$

from where it follows an affirmative answer for the corresponding 3SP problems. However, although

X has property (u) $\qquad\qquad\qquad$ $\Leftrightarrow DBSC(X) = B_1(X)$
X has Mazur property $\qquad\qquad\qquad$ $\Leftrightarrow X = S_1(X)$
X is a Grothendieck space $\qquad\qquad$ $\Leftrightarrow S_1(X) = X^{**}$

and, for separable spaces,

X contains no copy of l_1 $\qquad\qquad\qquad$ $\Leftrightarrow B_1(X) = X^{**}$

hold, those identifications do not directly provide 3SP results because the classes are not injective or surjective. In some cases, such as property (u), it is not a 3SP

property (see 6.9.b); in other, such as not containing l_1 or Grothendieck property, they are $3SP$ properties.

The difficult thing to obtain from a functor is, usually, the right-exactness. Results asserting that a functor is right exact are also known as "liftings".

2.4   Lifting of sequences

Let S: **Ban** → **Set** be a functor assigning to each Banach space a certain space SX of sequences in X having a certain property. For instance, consider the functor c assigning to a Banach space X the space $c(X)$ of convergent sequences of X , and to an operator T the induced operator between sequence spaces $(x_n) \to (Tx_n)$. Given a short exact sequence $0 \to Y \to X \to Z \to 0$, the sequence $0 \to c(Y) \to c(X) \to c(Z)$ is trivially exact. The problem is whether the last arrow is surjective. A result asserting that is usually called a "lifting result." In the case under consideration, what one needs is that "given a convergent sequence (z_n) of Z there is a convergent sequence (x_n) of X such that $qx_n=z_n$," and in general, something like "given a sequence (z_n) of type S in X/Y there is a sequence (x_n) of type S in X such that $qx_n=z_n$." That much cannot be, in general, obtained and it is necessary to impose some condition on the kernel of q.

Precisely, a lifting is a result having the form: if $Q: X \to Z$ is a quotient map and (z_n) is a sequence in Z having property P then, under some hypothesis on $Ker\ Q$, there is a sequence (x_n) in X having property P and such that $Qx_n=z_n$.

A classical result in this direction is:

Lohman's lifting 2.4.a. *Let $Q: X \to Z$ be a quotient map. If $KerQ$ contains no copy of l_1, every weakly Cauchy sequence (z_n) in Z admits a subsequence that is the image under the quotient map of a weakly Cauchy sequence in X.*

Proof. Let (x_n) be a bounded sequence in X without weakly Cauchy subsequences such that (Qx_n) is weakly Cauchy in Z. By Rosenthal's lemma, some subsequence (x_m) is equivalent to the canonical basis of l_1 and still is mapped to a weakly Cauchy sequence (Qx_m) of Z. Since the difference sequence $(Qx_{2n+1} - Qx_{2n})$ is weakly null, a certain sequence of convex combinations (σ_n) converges to 0 in norm in Z. If (μ_n) denotes the same convex combination of (x_m) it turns out that $Q\mu_n=\sigma_n$ although (μ_n) is still equivalent to the canonical basis of l_1. And if (y_n) is some sequence in $Ker\ Q$ such that

$$\lim \| \mu_n - y_n \| \le \lim 2 \| Q\mu_n \| = 0$$

some subsequence of it (y_j) can be chosen equivalent to (μ_j), in turn equivalent to the canonical basis of l_1. This contradiction proves the result. $\square$

Translated to the context of exact sequences, if *wcc* denotes the functor assigning to X the space $wcc(X)$ of weakly conditionally compact sequences (i.e., bounded sequences whose subsequences admit weakly Cauchy subsequences) then Lohman's lifting reads: *if* $0 \rightarrow Y \rightarrow X \rightarrow Z \rightarrow 0$ *is a short exact sequence where Y does not contain* l_1 *then the sequence* $0 \rightarrow wcc(Y) \rightarrow wcc(X) \rightarrow wcc(Z) \rightarrow 0$ *is exact.*

This allows one to give a functorial proof that *not containing* l_1 is a 3*SP* property: denote by b the functor assigning to a Banach space X the space $b(X)$ of bounded sequences of X. The functor b is exact. Consider the natural transformation $i: wcc \rightarrow b$ defined by canonical inclusion. Observe that i_X is surjective if and only if X does not contain l_1; and thus, by proposition 2.3.a, the class

$$\{X \in \textbf{Ban}: i_X: wcc(X) \rightarrow b(X) \text{ is surjective}\} = \{X \in \textbf{Ban}: X \text{ does not contain } l_1\}$$

is a 3*SP* ideal.

The following lifting results are in [148]:

Lifting (of weakly convergent sequences) 2.4.b. *Let* $Q: X \rightarrow Z$ *be a quotient map. If KerQ is reflexive then each bounded sequence in X whose image in Z is weakly convergent admits a weakly convergent subsequence.*

Proof. Let (x_n) be a bounded sequence in X with (Qx_n) weakly convergent to Qx in Z. If x^{**} is in the weak*-closure of $\{x_n\}$ in X^{**}, then $x^{**} - x \in (KerQ)^{**} = KerQ$. this shows that $x^{**} \in X$ and (x_n) admits a weakly convergent subsequence. $\square$

The immediate translation is the following. Let *Wc* the functor assigning to X the space $Wc(X)$ of relatively weakly compact sequences of X (i.e., sequences whose subsequences admit weakly convergent subsequences): if Y is reflexive, then *Wc* is exact. If $i: Wc \rightarrow b$ denotes the natural transformation given by canonical inclusion, and since i_X is surjective if and only if X is reflexive, by proposition 2.3.a,

$$\{X \in \textbf{Ban}: i_X: Wc(X) \rightarrow b(X) \text{ is surjective}\} = \{X \in \textbf{Ban}: X \text{ is reflexive}\}$$

is a 3*SP* ideal.

Lifting (of weakly convergent sequences) 2.4.c. *Let* $Q: X \rightarrow Z$ *be a quotient map. If KerQ is weakly sequentially complete and* (x_n) *is a weakly Cauchy sequence in X such that* (Qx_n) *is weakly convergent then* (x_n) *is weakly convergent.*

Proof. Let (x_n) be a weakly Cauchy sequence in X with (Qx_n) weakly convergent to Qx in Z. Let x^{**} be the weak*-limit in X^{**} of (x_n). Again, $x^{**}-x \in (KerQ)^{**}$. If Y is a subspace of a Banach space X, Baire 1-elements in X^{**} that belong to Y^{**} are Baire 1-elements of Y^{**} (see e.g. [89]); which means that $x^{**}-x$ is the weak* limit in $(KerQ)^{**}$ of a sequence of elements of $KerQ$. Since $KerQ$ is weakly sequentially complete, $x^{**}-x \in KerQ$ and $x^{**} \in X$. □

This lifting can also be translated into the functorial language. Denote by c^w and c^{wC} the functors assigning to X the spaces $c^w(X)$ of weakly convergent sequences and $c^{wC}(X)$ of weakly Cauchy sequences of X respectively. If $i: c^w \to c^{wC}$ denotes the natural transformation given by canonical inclusion, and taking into account that i_X is surjective if and only if X is weakly sequentially complete, one obtains that the class

$$\{X \in \textbf{Ban}: i_X: c^w(X) \to c^{wC}(X) \text{ is surjective}\}$$

$$= \{X \in \textbf{Ban}: X \text{ is weakly sequentially complete}\}$$

is a 3*SP* ideal.

The 3*SP* problem for weak sequential completeness has been considered by several authors and seems to be folklore: Jarchow [184] gives a proof using the Hahn-Banach theorem; Rakov [302] gives a proof based in l_1's presence, and Pisier [294] blames Godefroy to have priority in this matter.

Since the exactness of the sequence $0 \to Y \to X \to Z \to 0$ implies the exactness of the dual sequence $0 \to Z^* \to X^* \to Y^* \to 0$, it is possible to work with this sequence to obtain 3*SP* results for properties relating either weak* convergent and weakly convergent sequences; or weak* Cauchy and weakly Cauchy sequences. We shall consider Grothendieck property (weak* convergent sequences are weakly convergent) and not having quotients isomorphic to c_0 (weak* Cauchy sequences admit weakly Cauchy subsequences).

Lifting (of weakly Cauchy sequences in duals) 2.4.d. *Let $0 \to Y \to X \to Z \to 0$ be a short exact sequence. Assume that Z has no quotients isomorphic to c_0. If (f_n) is a weakly* convergent sequence in X^* such that (i^*f_n) is weakly Cauchy in Y^*, then (f_n) has a weakly Cauchy subsequence.*

Proof. The proof is similar in spirit to that of Lohman's lifting, but working with the dual sequence $0 \to Z^* \to X^* \to Y^* \to 0$. Assume that (f_n) has no weakly Cauchy subsequences. Some subsequence (let it be (f_n) itself) is thus equivalent to the canonical basis of l_1, and the same happens to $(g_n)=(f_{2n+1} - f_{2n})$. Since (i^*f_n) is weakly Cauchy, (i^*g_n) is weakly convergent to 0 and there is a convex combination

σ_n of the g_n such that $(i^*\sigma_n)$ is norm convergent to 0. The sequence (σ_n) is still equivalent to the canonical basis of l_1. Choosing $(h_n) \in Z^*$ such that $\| h_n - \sigma_n \| \leq 2^{-n}$ we obtain a sequence in Z^* that is equivalent to the canonical basis of l_1. But at the same time the sequence (h_n) is weak* null since so are (f_n) and (σ_n).

But a weak* null sequence (h_n) in Z^* equivalent to the canonical basis of l_1 induces a quotient map $Z \to c_0$. $\qquad\qquad\square$

Lifting (of weakly convergent sequences in duals) 2.4.e. *Let $0 \to Y \to X \to Z \to 0$ be a short exact sequence. Let Z be a Grothendieck space. If (f_n) is a weakly* convergent sequence in X^* such that (i^*f_n) is weakly convergent in Y^*, then (f_n) has a weakly convergent subsequence.*

Proof. Since Grothendieck spaces have no quotients isomorphic to c_0 and their dual spaces are weakly sequentially complete, by 2.4.d (f_n) has a weakly Cauchy subsequence, which is weakly convergent by 2.4.c. $\qquad\qquad\square$

It follows that the Grothendieck property and " to admit no quotient isomorphic to c_0 " are 3SP properties (see also 6.6.o).

The following lifting has an obvious proof. We include it for the sake of completeness.

Lifting of weak* convergent sequences 2.4.f. *Let $0 \to Y \to X \to Z \to 0$ be a short exact sequence. Let X have weak* sequentially compact dual ball. If (f_n) is a bounded sequence in X^* such that (i^*f_n) is weak* convergent to y^* then there is a subsequence (f_k) weak* convergent to some x^* with $i^*x^*=y^*$.*

Bounded basic sequences cannot be lifted. Terenzi shows in [347] the rather pathological behaviour of basic sequences by constructing (in the hardest meaning of the word) a short exact sequence $0 \to Y \to X \to Z \to 0$, where Z admits a bounded basis (z_n) in such a way that no bounded basic sequence (x_n) in X satisfies $qx_n=z_n$. This answers a problem posed by Bor-Luh Lin in Banff in 1988.

A second look at liftings

The following results are elementary. Our interest into these proofs is that they contain an essential ingredient common to most 3SP proofs: the existence of a *lifting* (i.e., a selection for the quotient map) preserving the property under consideration.

Recall that if X is a Banach space, $dens(X)$ is the smallest cardinal such that X admits a dense subset of that cardinal. X is separable if and only if $dens(X)=\omega$. The following result appears explicitly in Domanski [98] ("obvious proof")

Lemma 2.4.g. $dens(X)= max\{ dens(Y), dens(X/Y)\}$.

Proof. If $\{q(x_i)\}$ and $\{y_j\}$ are dense subsets of X/Y and of Y respectively, then $\{x_i + y_j\}$ is a dense subset of X. $\qquad\square$

In particular,

Theorem 2.4.h. *Separability is a 3SP property.*

The same argument, considering the dimension of the spaces, shows

Theorem 2.4.i. *To be finite dimensional is a 3SP property.*

which can also be proved via lifting of convergent sequences.

Thus, everything seems to depend on the existence of a "selection map" $X/Y \to X$. This selection is not a map of the category **Ban** (a section) since otherwise the short sequence would split. Lifting results imply the existence of a certain selection. We precise this: Let $0 \to Y \to X \to Z \to 0$ be an exact sequence and let F and G: **Ban** $\to$ **Set** be functors.

Proposition 2.4.j. *Assume that FZ is a subset of $FZ \cap GZ$ and that there is a selection s as in the diagram*

$$FX \to FZ$$
$$\subseteq$$
$$FX \cap GX \xleftarrow{s} FZ \cap GZ$$

Then $\{X \in \textbf{Ban}: FX \subset GX\}$ is a 3SP property.

Proof. Given an element $x \in FX$ then an element $sqx \in FX \cap GX$ can be assigned in such a way that $y=x-sqx \in FY \subset GY$. Thus $x-sqx \in GX$ and $x=sqx+y \in GX$.

As it should be by now clear, the core of any positive 3SP result is the existence of a selection (or lifting) for the quotient map $X \to X/Y$. The following types of selections are possible:

(C)   Continuous (non linear) homogeneous selection (Bartle-Graves, see 2.2.1).

(B)   Bounded (non linear, not continuous) homogeneous selection that can be easily obtained taking into account that the quotient map is open.

(L)   Linear (unbounded) selection obtained by taking a representative belonging
       to a fixed algebraic complement of the kernel.

2.5   Lifting for operators

Many properties of Banach spaces are described by the coincidence of two classes
of operators. Among these, one should distinguish between *covariant* and
contravariant properties.

Covariant properties are those defined by the coincidence of two operator ideals
A and B in the form: the space X has the property BA^{-1} if for all spaces M, A(M,
X) $\subset$ B(M, X); these shall be treated in 6.7. Observe that the functors there
considered A(M, $\cdot$) and B(M, $\cdot$) are covariant.

Contravariant properties are more commonly used. They are defined by the
coincidence of two operator ideals A and B: a space X has property $A^{-1}B$ if for all
spaces M, A(X, M) $\subset$ B(X, M). Observe that the functors here considered A($\cdot$, M)
and B($\cdot$, M) are contravariant. Contravariant properties are considered in Chapter
6 (see also 2.7).

Let P be the property: for all spaces M, A(X, M) $\subset$ B(X, M). One way to prove
that P is a 3*SP* property is to (try to) apply the "exact functor" schema as follows:
consider $0 \to Y \to X \to Z \to 0$ a short exact sequence where both Y and Z have P.
Given a Banach space M, apply the contravariant functors A($\cdot$, M) and B($\cdot$, M)
obtaining (and this is the first problem) short exact sequences. Now, a natural
transformation (whose existence is the second difficulty) τ: A($.$, M) $\to$ B($.$, M) and
the three-lemma would finish the proof.

The existence of τ seems to be the major difficulty. Indeed, although the arrows
τY and $\tau X/Y$ exist, an arrow τX might not exist. In other words, maybe no natural
transformation τ can be defined. Still, the approach of proposition 2.4.j could work,
and sometimes does.

Trying to obtain the exactness of the functors considered is what we will do
now. The contravariant functor $Hom(\cdot, M)$ is left exact, i.e., if $0 \to Y \to X \to Z$
$\to 0$ is a short exact sequence then

$$0 \to Hom(Z, M) \to Hom(X, M) \to Hom(Y, M)$$

is exact, and there is no guarantee that the last arrow is surjective. When A is an operator ideal, there are two obstacles to obtain the exactness of the sequence

$$0 \to A(Z, M) \to A(X, M) \to A(Y, M) \to 0,$$

namely, the surjectivity of the last arrow and the exactness at the middle point.

The surjectivity of $A(i, M)$: $A(X, M) \to A(Y, M)$ amounts the existence of an extension for operators in A, i.e., that given an operator t: $Y \to M$ in A, there is some operator T: $X \to M$ also in A and such that $Ti=t$. In the next section 2.6 we shall give a certain context for extensions and liftings for operators. In some cases such as

$$0 \to Y \to Y^{**} \to Y^{**}/Y \to 0$$

this selection $t \to t^{**}$ exists (e.g, $A = W$) and for this reason $3SP$ results are sometimes available for the couple Y, Y^{**}/Y when no general $3SP$ result exist. See results in 6.6 where it is shown that the Dunford-Pettis property $(W \subset C_\infty)$ is not a $3SP$ property; this means that there is no possibility for a selection

$$C_\infty(Y, M) \to C_\infty(X, M) \cap W(X, M).$$

Nevertheless, when $X=Y^{**}$ and Y does not contain l_1 (completely continuous operators are compact) then such a selection exists: $t \to t^{**}$ and thus a positive $3SP$ result is available.

The exactness of the functors at the middle point can be presented in an elegant form. Let A be an operator ideal, M a Banach space and $0 \to Y \to X \to Z \to 0$ a short exact sequence. The exactness of the sequence

$$A(Z, M) \to A(X, M) \to A(Y, M)$$

means that $Im\ A(q, M) = Ker\ A(i, M)$. Elements in $Im\ A(q, M)$ are arrows having the form Tq, where $T \in A(Z, M)$. They necessarily belong to $Ker\ A(i, M)$ since $Tqi=0$. All this means that $Im\ A(q, M) \subset Ker\ A(i, M)$. The other containment is different. What one needs is that if $T \in A(X, M)$ is such that $Ti=0$ then T factorizes through q in the form $T=Rq$ with $R \in A(Z, M)$. Certainly, T admits a factorization Rq; what is not clear is whether one can obtain $R \in A$.

Recall that given an operator ideal A the surjective hull is defined as the operator ideal $A^{sur} = \{T$: for some quotient map q, $Tq \in A$ $\}$. The operator ideal A is said to be surjective if $A = A^{sur}$. We have proved that if A is a surjective operator ideal, then the exactness is preserved at the middle point. If, moreover, there is some extension result for A one should have an exact functor.

An interesting case where the situations described appear is when considering the operator ideal Π_1 of 1-summing operators. Recall that an operator $T: X \to Y$ is 1-summing if it transforms weakly unconditionally summable series into absolutely summable series. It turns out is the following.

Proposition 2.5.a. *Let* $i: Y \to X$ *be an into isomorphism, and let M be a Banach space. Then the map* $T \to Ti$ *from* $\Pi_1(X, M)$ *into* $\Pi_1(Y, M)$ *is a quotient map.*

Proof. The Grothendieck-Pietsch factorization theorem (see [90] for a display of the theorem in its full glory) provides factorization of 1-summing operators through some L_∞-space. Now, if $t: Y \to M$ is a 1-summing operator then t factorizes as ab where $b: Y \to L_\infty$. This operator b has an extension $B: X \to L_\infty$ in such a way that $T=aB: X \to M$ extends t and is 1-summing. $\square$

Many more results on lifting of operators can be found in [333].

2.6   Three-space properties and semigroups of operators

Let A be an operator ideal. Let us consider some semigroups A_+ and A_- associated to A. In what follows, T, R and S shall denote continuous operators acting on the appropriate spaces:

$$T \in A_+ \quad \text{if and only if } TS \in A \text{ implies } S \in A.$$
$$T \in A_- \quad \text{if and only if } RT \in A \text{ implies } R \in A.$$

For instance, if K denotes the ideal of compact operators, K_+ are the *upper semi-Fredholm* operators while K_- are the *lower semi-Fredholm* operators. For W the ideal of weakly compact operators, W_+ and W_- are the *tauberian* and *cotauberian* operators, respectively. Those semigroups of operators, and others (associated to Rosenthal, Dunford-Pettis and Dieudonné operators) appear studied in [150].

We shall say that an operator ideal A is *preserved by liftings* if quotient maps with kernel in $Sp(A)$ belong to A_+; i.e., if q is surjective, $Ker\, q \in Sp(A)$ and qS belongs to A then also S belongs to A.

Theorem 2.6.a. *Let A be an operator ideal preserved by liftings. Then* $Sp(A)$ *is a 3SP space ideal.*

Proof. Let $0 \to Y \to X \to Z \to 0$ be a short exact sequence where both Y and Z belong to $Sp(A)$. Since $q \in A_+$ and $q\, id_X = id_Z\, q \in A$ it follows that $id_X \in A$ $\square$

Examples. Observe that Lohman's lifting can be expressed by saying that the operator ideal of Rosenthal's operators (those transforming bounded sequences into sequences admitting weakly Cauchy subsequences) is preserved by liftings. Other operators that are preserved by liftings are: compact, weakly compact, C_p for $1 \leq p \leq \infty$, and weakly completely continuous (or Dieudonné) operators.

Dually, we say that an operator ideal A is preserved by extensions if isomorphisms into with cokernel in $Sp(A)$ belong to A_+; i.e., if Y is a subspace of X with injection $i: Y \to X$ and $X/Y \in Sp(A)$ then $Ri \in A$ implies that $R \in A$.

Theorem 2.6.b. *Let* A *be an operator ideal preserved by extensions. Then* $Sp(A)$ *is a 3SP space ideal.*

Proof. Let $0 \to Y \to X \to Z \to 0$ be a short exact sequence where both Y and Z belong to $Sp(A)$. Since $i \in A_-$ and $id_X\, i = i\, id_Y \in A$ it follows that $id_X \in A$. $\square$

Proposition 2.6.c. *If* A *is preserved by liftings then* A^{dual} *is preserved by extensions. If* A *is preserved by extensions then* A^{dual} *is preserved by liftings.*

The proof is trivial.

A sophisticated form of lifting was presented by Jarchow in [184]:

2.7   Jarchow's factorization method

Observe that properties of a Banach space X defined by the coincidence, for every space Y, of two classes of operators $A(X, Y) \subset B(X, Y)$, where A and B are two given operator ideals, can be reformulated as $Id_X \in A^{-1} \circ B$, where

$$T \in A^{-1} \circ B(X, Y) \Leftrightarrow \forall S \in A(Y, Z),\ ST \in B(X, Z).$$

Recall that given operator ideals A and B, $A \circ B = \{ST: S \in A \text{ and } T \in B\}$.

For $1 \leq p \leq \infty$, F_p denotes the ideal of operators that are factorizable through l_p. The paper [184] contains the next result:

Proposition 2.7.a. *Let* A *and* B *be operator ideals such that* A *is injective and* B *satisfies* $F_1 \circ B \subset A$. *Then* $Sp([F_1 \circ B] \circ A^{-1})$ *and* $Sp([F_1 \circ B]^{sur} \circ A^{-1})$ *are 3SP ideals.*

Proof. Let $0 \to Y \to X \to Z \to 0$ be a short exact sequence, and assume that both id_Y and id_Z belong to $[F_1 \circ B] \circ A^{-1}$. Given $S \in A(Z, X)$, the hypothesis on Z implies $qS \in F_1 \circ B(M, Z)$. By the lifting property of $l_1(\Gamma)$-spaces, qS can be lifted to $l_1(\Gamma)$ and thus there is some $T \in F_1 \circ B(M, X)$ such that $qT = qS$. Since $q(T-S) = 0$, the operator $T--S$ takes its values in Y. Since $F_1 \circ B \subset A$, $T-S \in A(M, X)$, and since

A is injective, $T-S \in A(M, Y) \subset F_1 \circ B(M, Y)$; hence $T-S \in F_1 \circ B(M, X)$ and since $T \in F_1 \circ B(M, X)$ also $S \in F_1 \circ B(M, X)$. The ideal $[F_1 \circ B]^{sur} \circ A^{-1}$ can be treated similarly.   □

By duality, the extension property of $l_\infty(\Gamma)$-spaces yields:

Proposition 2.7.b. *Let* A *and* B *be operator ideals such that* A *is surjective and* B *satisfies* $B \circ F_\infty \subset A$. *Then* $Sp(A^{-1} \circ [B \circ F_\infty])$ *and* $Sp(A^{-1} \circ [B \circ F_\infty]^{inj})$ *are* 3*SP ideals.*

2.8   Incomparability classes

A general method was developed in [7] to obtain 3*SP* ideals starting from *any* class Λ of Banach spaces. Given such a class Λ, two space ideals $\Lambda(i)$ and $\Lambda(c)$ are defined as follows:

$$X \in \Lambda(i) \Leftrightarrow X \text{ is incomparable with all spaces } Z \in \Lambda,$$
$$X \in \Lambda(c) \Leftrightarrow X \text{ is co-incomparable with all spaces } Z \in \Lambda.$$

Recall that two Banach spaces X and Z are said to be *incomparable* if no infinite dimensional subspace of X is isomorphic to a subspace of Y. The spaces X and Y are termed *co-incomparable* if there is no infinite dimensional quotient of X isomorphic to a quotient of Y. One has [7]:

Theorem 2.8.a. *For any class* Λ *of Banach spaces, the classes* $\Lambda(i)$ *and* $\Lambda(c)$ *are* 3*SP ideals.*

Proof. Only the 3*SP* property is not trivial. Let $X \notin \Lambda(i)$ and $Y \subset X$ such that $Y \in \Lambda(i)$. There exists an infinite dimensional subspace N of X isomorphic to a subspace of a space in Λ. By [309], or else [7, 2.3 (*i*)], $Y+N$ is closed and $Y \cap N$ finite-dimensional; obviously, we may suppose $Y \cap N = \{0\}$ without restricting the generality. Then $q(N)$ is a subspace of X/Y isomorphic to N, and so $X/Y \notin \Lambda(i)$. Therefore, $Y, X/Y \in \Lambda(i)$ implies $X \in \Lambda(i)$.

The proof for $\Lambda(c)$ is quite analogous, taking [7, 2.3.(*ii*)] into account, where it is proved that if X/M and X/N are totally coincomparable then $M+N$ is closed.   □

Observe that the procedures can be iterated to define $\Lambda(ii)$ and $\Lambda(cc)$. However, $\Lambda(iii) = \Lambda(i)$ and $\Lambda(ccc) = \Lambda(c)$.

Examples. Proposition 2.8.a implies that given *any* class Λ the classes $\Lambda(i)$, $\Lambda(ii)$, $\Lambda(c)$ and $\Lambda(cc)$ are 3*SP* ideals. Interesting examples are then obtained starting with a *minimal* Banach space E, i.e. a Banach space such that every infinite dimensional subspace contains an isomorphic copy of E (see [6]). The spaces l_p, $1 \leq p < +\infty$, c_0, Tsirelson space and Schlumprecht distortable space [322] are examples of minimal spaces. If E is a minimal space and one considers the class $\{E\}$ then

$$\{E\}(i) = \{spaces\ not\ containing\ copy\ of\ E\},$$
$$\{E\}(ii) = \{spaces\ which\ are\ hereditarily\ E\},$$

are $3SP$ ideals.

Analogously, a Banach space E is said to be *co-minimal* if every infinite dimensional quotient of E has a quotient isomorphic to E. Examples of co-minimal spaces are l_p, $1 < p < \infty$, c_0 and Tsirelson's dual space. For a co-minimal space E the classes $\{E\}(c)$ and $\{E\}(cc)$ admit a natural description. More examples:

If $R = \{reflexive\ spaces\}$ then

$$R(i) = \{spaces\ containing\ no\ reflexive\ subspaces\},$$
$$R(ii) = \{somewhat\ reflexive\ spaces\},$$

are $3SP$ space ideals. If $U = \{spaces\ having\ unconditional\ basis\}$ then, taking into account that blocks of unconditional basis are unconditional basic sequences, then

$U(i) = \{spaces\ without\ unconditional\ basic\ sequences\},$
$U(ii) = \{spaces\ such\ that\ every\ subspace\ has\ an\ unconditional\ basic\ sequence\},$

are $3SP$ ideals. Observe that the example of Gowers and Maurey [156] shows that the classes $U(i)$ and $U(ii)$ are not trivial.

A Banach space X is said to be *indecomposable* if there is no decomposition $X = M \oplus N$ with M and N closed infinite dimensional subspaces. The space X is said to be *hereditarily indecomposable* (H.I.) if every infinite dimensional subspace is indecomposable. Recent work of Gowers and Maurey [156] shows that there exist H.I. spaces. Moreover, Gowers [155] proves that given a Banach space X then it contains either a H.I. subspace or a subspace with an unconditional basis. It is then clear that

$$\text{H.I.}(i) = U(ii)$$
$$\text{H.I.}(ii) = U(i).$$

Question. *Is H.I. a 3SP property?*

In addition to this, it is worth to remark that Odell [268] defines property WU as: every normalized weakly null sequence admits an unconditional subsequence.

Question. *Is WU a 3SP property?*

2.9   *P-by-Q* and **P⊕Q** properties

In group theory it is said that a group G is a *P-by-Q* group if there is some normal
subgroup N such that N has property P and G/N has Q. We shall use this
terminology; thus, given two properties P and Q, we will say that a Banach space
X has the *P-by-Q property* if it admits a closed subspace Y such that Y has property
P and X/Y has Q. In this context, P is a 3SP property if and only if *P-by-P*
implies P.

Lemma 2.9.a. *Let* $0 \rightarrow Y \rightarrow X \rightarrow Z \rightarrow 0$ *be a short exact sequence with Z separable.
There is a separable subspace M of X such that* $Y+M=X$.

Proof. Let $B: Z \twoheadrightarrow X$ be a bounded selection for the quotient map. Take a dense
sequence (z_n) in the unit ball of Z. Let M be the closed span of $\{Bz_n\}_{n \in \mathbf{N}}$. It is
clear that for some $r>0$ $q(rBall(M))$ is dense in $Ball(Z)$; this makes $q(M)=Z$, and
thus $M+Y=X$.   □

This result justifies the hypotheses of the following theorem.

Theorem 2.9.b. *Let P and Q be* 3SP *properties. Assume that Q passes to quotients
and P satisfies the additional condition: if X/Y has P then there is some subspace
N of X having P such that* $N+Y=X$. *Then P-by-Q is a* 3SP *property.*

Proof. Let $0 \rightarrow Y \rightarrow X \rightarrow Z \rightarrow 0$ be a short exact sequence where both Y and Z have
the *P-by-Q* property. Since Y has the *P-by-Q* property, there is a subspace M of Y
having property P such that Y/M has property Q. Since $Z=X/Y$ has *P-by-Q* property
there is a subspace V of X containing Y such that V/Y (which should be thought as
subspace of X/Y) has property P and $(X/Y)/(V/Y) = X/V$ has property Q. Since Y/M
is a subspace of V/M and $(V/M)/(Y/M) = V/Y$ has property P, there is some
subspace F/M of V/M having property P and such that $F/M + Y/M = V/M$. Both
M and F/M have property P and therefore F has property P. It only remains to
show that X/F has property Q. Since there is an exact sequence $0 \rightarrow V/F \rightarrow X/F \rightarrow$
$X/V \rightarrow 0$ and we already know that X/V has Q, it is only necessary to verify that
V/F has Q. This is a consequence of

$$V/F = (F+Y)/F = Y/(Y \cap F) = (Y/M)/((Y \cap F)/M).$$ □

Theorem 2.9.c. *Let P and Q be* 3SP *properties. Assume that P-by-Q implies Q-by-
P. Then Q-by-P is a* 3SP *property.*

Proof. To clarify the proof, we state two claims and then the proof of the theorem.
Let $0 \rightarrow Y \rightarrow X \rightarrow Z \rightarrow 0$ be a short exact sequence.

Claim. If Y has P-by-Q and Z has Q then X has P-by-Q.

Proof of the claim. Let $0 \to N \to Y \to Y/N \to 0$ be a short exact sequence where N has P and Y/N has Q. The sequence $0 \to Y/N \to X/N \to Z \to 0$ is also exact, and since Q is a $3SP$ property, X/N has Q; thus X has P-by-Q.

Analogously, *If Y has P and Z has P-by-Q then X has P-by-Q.*

Proof of the theorem. Since Y has Q-by-P, let $0 \to N \to Y \to Y/N \to 0$ be a short exact sequence where N has Q and Y/N has P. Since Z has Q-by-P, let M be a subspace of X containing Y such that $0 \to M/Y \to X/Y \to X/M \to 0$ is a short exact sequence, M/Y has Q and X/M has P.

The sequence $0 \to Y/N \to M/N \to M/Y \to 0$ is exact, so M/N has P-by-Q and, therefore, Q-by-P. The sequence $0 \to M/N \to X/N \to X/M \to 0$ is exact; and since M/N has Q-by-P and X/M has P, by the first claim X/N has Q-by-P. Finally, the sequence $0 \to N \to X \to X/N \to 0$ is exact; and since N has Q and X/N has Q-by-P, by the second claim X has Q-by-P.   □

Observe that the preceding proof could have been simplified after

The -by- observation 2.9.d. *Operation -by- is associative!*

whose proof is a simple pull-back and push-out argument.

Given two properties P and Q, we will say that a Banach space X has the $P \oplus Q$ property if it can be decomposed as $X = Y \oplus Z$ where Y has property P and Z has Q.

Question. *When P-by-Q implies $P \oplus Q$?*

Nevertheless, a proper reading of theorem 2.9.c says when $P \oplus Q$ is a $3SP$ property, at least when $\{0\}$ has P and Q.

Theorem 2.9.e. *Let P and Q be $3SP$ properties shared by $\{0\}$. Then $P \oplus Q$ is a $3SP$ property if and only if $P \oplus Q = P$-by-$Q = Q$-by-P.*

For instance, since

Proposition 2.9.f. *Reflexive-by-Separable implies Reflexive $\oplus$ Separable.*

Proof. Let $0 \to Y \to X \to Z \to 0$ be a short exact sequence with Y reflexive and Z separable. By lemma 2.9.a, $X = M + Y$ for some separable subspace M of X. Let (m_n) be a null total sequence in M. The set $\{m_n\}_n + Ball(Y)$ is weakly compact in X and spans a dense subspace; i.e., X is *weakly compactly generated* (see 4.10) and thus M is contained in a complemented separable subspace N whose complement is necessarily reflexive since it is a quotient of Y.   □

One has.

Theorem 2.9.g. *Separable-by-Reflexive is a* 3*SP property.*

and since *Separable-by-Reflexive* does not imply *Separable* $\oplus$ *Reflexive* (by the Johnson-Lindenstrauss example, to be studied in 4.10, it follows that *Reflexive-by-Separable* = *Reflexive* $\oplus$ *Separable* is not a 3*SP* property.

For more applications, recall that l_1 is projective and l_∞ is injective. Thus,

Proposition 2.9.h. *Let P be a* 3*SP property. Then P-by-*$\{l_\infty\}$ *and* $\{l_1\}$*-by-P are* 3*SP properties.*

Proof. Since l_∞ is injective, $\{l_\infty\}$*-by-P* implies $l_\infty \oplus P$. Since l_1 is injective, *P-by-*$\{l_1\}$ implies $P \oplus \{l_1\}$ $\square$

Analogously, since c_0 is separably injective

Proposition 2.9.i. *Separable-by-*$\{c_0\}$ *is a* 3*SP property.*

An old result of Lindenstrauss [243] shows that any exact sequence $0 \to Y \to X \to Z \to 0$ where Y is complemented in its bidual and Z is an $\mathcal{L}_1$-space splits. Therefore

Proposition 2.9.j. *Let P be a* 3*SP property such that spaces in P are complemented in its bidual. Then* $\mathcal{L}_1$*-by-P is a* 3*SP property.*

Question. *Are* $\mathcal{L}_1$*-by-*$\mathcal{L}_\infty$ *or* $\mathcal{L}_1$*-by-separable* 3*SP properties?*

Appendix 2.10

Applications of 3SP results to the opening

The gap (or opening) between subspaces

The gap between two (closed) subspaces of a Banach space was introduced to study the perturbation of closed semi-Fredholm operators. The book [218] and the survey [275] are good references for this topic. If T and $T+K$ are closed operators, the gap between the graphs $G(T)$ and $G(T+K)$ gives an estimation of the "size" of the perturbation produced by K on the operator T.

Several authors have analyzed the stability of some isomorphic properties of Banach spaces under the gap in the following sense: Assume that M is a (closed) subspace of a Banach space which enjoys some isomorphic property P. Is it possible to find $r>0$ so that if the gap from M to another subspace N is less than r, then N enjoys P? Positive answers have been obtained for several properties such as separability, reflexivity, containing no copies of l_1, etc. Moreover, it has turned out that a negative answer for property P is related with the failure of the $3SP$ property for P. In fact, one has

If an isomorphic property is preserved by finite products and complemented subspaces, but fails the 3SP property, then this property is not stable under the gap.

In this section we describe the main results and problems about stability of properties under the gap. Let us recall some definitions. The *non-symmetric gap* $\gamma(M, N)$ between two subspaces M and N of a Banach space is given by

$$\gamma(M, N) = sup \{ dist(m, N) : m \in M, \|m\| = 1 \},$$

and the (symmetric) *gap* between M and N is given by

$$g(M, N) = max \{ \gamma(M, N), \gamma(N, M) \}.$$

Observe that g induces a complete metric in the class $G(X)$ of all closed subspaces of a Banach space X (see [275, Thm. 3.4 (g)].

Let Λ be a class of Banach spaces which is preserved by isomorphisms, and let $\delta > 0$. The class Λ is δ-*stable* if there exists a number $\delta > 0$ so that given a Banach space X and subspaces $M, N \in G(X)$, if $N \in \Lambda$ and $g(M, N) < \delta$, then $N \in \Lambda$. The class Λ is δ-*costable* if there exists a number $\delta > 0$ so that given a Banach space X and subspaces $M, N \in G(X)$, if $X/N \in \Lambda$ and $g(M, N) < \delta$, then $X/N \in \Lambda$. We say that Λ is *stable* (*costable*) if it has those properties for some number $\delta > 0$.

These concepts have been studied by Janz [179]. Analogously, it could be said that a property P is *open* if for every Banach space X the subclass

$$\{M \in G(X) : M \text{ has } P \}$$

is open in $(G(X), g)$. Observe that a stable class is open. That not every open class has to be stable is proved in Lemma 2.10.b.

Theorem 2.10.a. *Let P be a property that is preserved by products. If P is open then it is a 3SP property.*

Proof. Assume that P is not a 3SP property and let X be a Banach space without property P for which some subspace Y and the quotient space X/Y enjoy property P. Since P is preserved by products, $Y \times X/Y$ must have property P. Since

$$X_\varepsilon = \{(\varepsilon x, qx) : x \in X\}$$

is isomorphic to X, this space does not have property P. We claim that

$$g(Y \times X/Y , X_\varepsilon) \le \varepsilon,$$

thus proving that P is not open:

if $\| (y, qz) \| = 1$ in $Y \times X/Y$ then some element $y_z \in Y$ exists such that $\| z - y_z \| < 1 + \varepsilon$. If $x = z - y_z + y/\varepsilon$ then $\| (y, qz) - (\varepsilon x, qx) \| \le \varepsilon$. On the other hand if $\| (\varepsilon x, qx) \| = 1$ there is some element $y_x \in Y$ such that $\| x - y_x \| < 1 + \varepsilon$, and thus the choice $y = \varepsilon y_x$ and $z = qx$ verifies $\| (y, z) - (\varepsilon x, qx) \| \le \varepsilon$. □

The converse implication fails since 3SP properties need not be open: for example, the class $\{l_2\}(ii)$ of l_2-saturated Banach spaces has the 3SP property although it is not open (see 2.10.d).

Lemma 2.10.b. *The class of Banach spaces isomorphic to l_1 is open but not stable.*

Proof.

Not stable: Let $d: l_1 \to l_1$ be the dilation given by $d(x) = \varepsilon x$ and let $q: l_1 \to c_0$ be a quotient map. As in the preceding proof, $X_\varepsilon = \{(\varepsilon x, qx) : x \in X\}$ is isomorphic to l_1 while $g(Ker\ q \times c_0, X_\varepsilon) \le \varepsilon$.

Open: Assume that $M = [x_n]$ is a subspace of X such that $\| \Sigma a_n x_n \| \ge c \Sigma |a_n|$, where we assume $0 < c < 1$ and $\| x_n \| = 1$. Let us see that if N is a subspace of X such that $g(M, N) < c/5$ then N is isomorphic to l_1. To prove this, take some sequence (y_n) in N such that $\| x_n - y_n \| < c/5$. This spans l_1 in N since

$$\| \Sigma a_n y_n \| \quad \ge \quad \| \Sigma a_n x_n \| - \| \Sigma a_n (x_i - y_n) \|$$

$$\ge (4/5) \Sigma |a_n|.$$

To conclude, we show that $N = [y_n]$. To do that it is enough to see that if $z \in N$ and $\| z \| = 1$ then $dist(z, [y_n]) < 3/5$: take a vector $x \in M$ such that $\| z - x \| < c/5$, and assume $x = \Sigma b_n x_n$. One has

$$\| z - \Sigma b_n y_n \| \quad \le \quad \| z - \Sigma b_n x_n \| + \| \Sigma b_n (x_i - y_n) \|$$

$$\le c/5 + (c/5) \Sigma |b_n|$$

$$\le c/5 + (c/5)c^{-1} \| x \|$$

$$< c/5 + (1/5)(1+c/5)$$

$$< 3/5. \qquad \qquad \Box$$

Alternatively, the stability could have been defined in terms of the non-symmetric gap γ instead of g. In this case, since $\gamma(M, N) = 0$ implies that M is a subspace of N and X/N is a quotient of X/M, stable (co-stable) classes would be preserved by subspaces (quotients).

Since $\gamma(M, N) = \gamma(N^\circ, M^\circ)$ (see [218, IV.2.9]) and $(X/M)^* = M^\circ$, it is immediate that if P is δ-stable (δ-costable), then the dual class $P^d = \{X: X^* \in P\}$ is δ-costable (δ-stable).

Examples 2.10.c. *Let us show some examples of δ-stable and δ-costable classes.*

i)   *The class of finite dimensional spaces is 1-stable.*

ii)   *The classes: superreflexive spaces; spaces not containing copies of l_1; and convex Banach spaces are 1-stable.*

iii)   *The class of separable spaces is 1/2-stable and 1/2 costable.*

Proof of i) If M, N are closed subspaces of X, then $\gamma(M, N) < 1$ and [218, IV.2.6] imply dim $M \le$ dim N. $\qquad \qquad \Box$

Proof of ii) We give the proof for the class of Banach spaces containing no copies of l_1. Let M, N be closed subspaces of a Banach space X, and suppose that M contains a copy of l_1 and $\gamma(M, N) = \delta < 1$. By a result of James [175], M contains an almost isometric copy of l_1. Thus, taking $\alpha > 0$ such that $\delta < 1 - 2\alpha$, it is possible to find a normalized sequence (x_n) in M so that for any scalars $a_1, \ldots, a_n$

$$\| a_1 x_1 + \ldots + a_n x_n \| \geq (1-\alpha)(|a_1| + \ldots + |a_n|).$$

Pick elements $y_i \in N$ such that $\|x_i - y_i\| < 1 - 2\alpha$ for $i = 1, \ldots, n$; one has

$$\| a_1 y_1 + \ldots + a_n y_n \| \geq \alpha(|a_1| + \ldots + |a_n|)$$

for all scalars $a_1, \ldots, a_n$. Hence N contains a copy of l_1.   □

The proof for the classes of superreflexive and B-convex spaces is analogous, using characterizations given in [176] and [130], respectively.

Proof of iii) Using a density argument, it is easy to prove that the class of separable spaces is 1/2-stable. On the other hand, suppose X/M is separable and let $\varepsilon > 0$ such that $\gamma(M, N) = \delta < 1/2 - \varepsilon$.

We shall construct a separable subspace Y of X such that every $x \in X$ with $\|x\| = 1$ admits a decomposition $x = m + z$ with $y \in Y$, $z \in N$, and $\|z\| < 1 + \varepsilon$.

To this end, let (y_n) be a dense sequence in the unit sphere of X/M. Denoting by $q: X \to X/M$ the quotient map, it is possible to select a sequence (x_n) in X such that $y_n = qx_n$ and $\|x_n\| < 1 + \varepsilon/2$. Now, given $x \in X$ with $\|x\| = 1$ there exist $0 < t_1 \leq 1$ and n_1 such that

$$\| qx - t_1 y_{n_1} \| < \epsilon / 4^2.$$

Thus, there are $0 < t_2 \leq \varepsilon/4$ and n_2 such that

$$\| qx - t_1 y_{n_1} - t_2 y_{n_2} \| < \epsilon / 4^3.$$

In this manner, the series $\sum t_i y_{n_i}$ converges to qx, and taking as Y the subspace spanned by the sequence (x_n), one has

$$x - \sum t_i x_{n_i} \in M \text{ and } \| \sum t_i x_{n_i} \| < 1 + \epsilon.$$

Now, if X/N is not separable, then $N + Y$ is not dense in X, and there must exist some norm one element $x \in X$ such that $dist(x, N+Y) > 1 - \varepsilon$.

Write $x = y + z$, with $y \in Y$, $z \in N$ and $\|z\| < 1 + \varepsilon$. Choosing $n \in N$ such that

$$\| y \| y \|^{-1} - n \| + \varepsilon < 1/2$$

one deduces

$$\big|\, \|x - z\| - \|y\|\, n\, \big| < 1-\varepsilon$$

in contradiction with $dist(x, N+Y) > 1-\varepsilon$. Consequently X/N is separable. □

The following result was obtained by Kadets [198]. It could also be derived from the fact that the class of Hilbert spaces fails the 3SP property.

Proposition 2.10.d. *The class of all Hilbert spaces is not open.*

Sketch of Proof. It is possible to construct a Banach space X and closed subspaces M and M_n of X so that M is isomorphic to l_2, while M_n is isomorphic to l_{p_n} for some sequence $p_n \to 2$, and $g(M_n, M) \to 0$. □

Ostrovskii [273] has been able to adapt this construction to obtain

Proposition 2.10.e. *For $1 < p < \infty$, the class of all spaces isomorphic to l_p is not open.*

It is not difficult to prove that given a Banach space N that is either injective or projective, the class of all spaces isomorphic to N is stable. Analogously, the class of all spaces isomorphic to c_0 is stable [179]. Janz [179] conjectures that these are the only stable Banach spaces.

Proposition 2.10.f. *The classes of all $\mathcal{L}_1$-spaces and $\mathcal{L}_\infty$-spaces are stable.*

The following result establishes a rather surprising fact: a property is stable if and only if it is costable.

Proposition 2.10.g. *If P is δ-costable then P is $\delta/2$-stable. If P is δ-stable then P is $\delta/2$-costable.*

Proof. Let X be a Banach space, and M, N closed subspaces of X for which $g(M, N) = \delta > 0$. To prove the first assertion, let $E(M, N)$ be the product space $M \times N$ endowed with the norm

$$\|(x, y)\|_E = \max\{\|x\|, \|y\|, \delta^{-1}\|x - y\|\}.$$

We shall denote it simply by E. Denote by M_0 and N_0 the spaces M and N seen as subspaces of E. It is enough to show that M is isometric to E/N_0, N is isometric to E/M_0 and $g(M, N) \le 2\delta$.

Observe that if $x \in M$, $\|x\| = 1$, then $inf\{\delta^{-1}\|x - y\| : y \in N\} \le 1$. Hence, $\|(x, 0) + N\|_E = 1$ and the natural map $M \to E/N_0$ is an isometry. Analogously for N and E/M_0. Moreover, $\|(x, 0)\|_E = 1$ implies $\|x\| = \delta$. Thus,

$$inf\{\|x - y\| : y \in N\} \le \delta^2.$$

Also, $\| x \| = \delta$ and $\| x - y \| \leq \delta^2$ imply $\| y \| \leq \delta + \delta^2$. One thus obtains

$$\gamma(M_0, N_0) \leq \delta + \delta^2 \leq 2\delta,$$

and analogously for $\gamma(N_0, M_0)$.

To prove the second assertion, observe that the space $E(M^\circ, N^\circ)$ is a dual space. In fact, $E(M^\circ, N^\circ)$ is isometric to the dual of $F(X/M, X/N)$, which is the product space $X/M \times X/N$ endowed with the norm whose unit ball is the absolutely convex hull of the set

$$Ball(X/M) \cup Ball\ (X/N) \cup \{\delta^{-1}\ (z+M, -z+N) : z \in Ball(X) \}.$$

If A_0 and B_0 denote the spaces X/M and X/N seen as subspaces of F, it is enough to show that A_0 is isometric to X/M, B_0 is isometric to X/N and $g(A_0, B_0) \leq 2\delta.\ \square$

Not all 3SP properties give stable classes. The following classes are not stable, although all of them are 3SP ideals:

spaces containing no copies of c_0,
weakly sequentially complete spaces,
spaces with the Radon-Nikodym property.

That they are not stable can be seen as a consequence of the fact that l_1 enjoys those properties and has a quotient isomorphic to c_0 (see the proof of lemma 2.10.b).

Question. *Are they open?*

Unstability is related with the existence of uncomplemented subspaces, as can be seen in the following result.

Proposition 2.10.h. *Let M and N be subspaces of X, N complemented; and let π denote a continuous projection onto N. If $\gamma(M, N) < \| I - \pi \|^{-1}$, then*

   i) M is isomorphic to a subspace of N, and
   ii) X/N is isomorphic to a subspace of X/M.

Proof. Given $x \in M$, for every $n \in N$ one has

$$\| \pi x \| \quad \geq \quad \| x \| - \| (I - \pi)(x + n) \|.$$

Then

$$\| \pi x \| \quad \geq \quad \| x \| - \| I - \pi \|\ dist(x, N)$$

$$\geq\ (1 - \| I - \pi \|\ \gamma(M, N)\ ) \| x \|,$$

hence the restriction of π to M defines an isomorphism into a subspace of N.

The above inequality shows that $M \cap Ker\pi = \{0\}$ and that $M + Ker\pi = \pi^{-1}\pi M$ is closed; hence $X/N = Ker\pi$ is also isomorphic to a subspace of X/M. $\square$

Identifying a Banach space with its natural isometric copy in the bidual (when necessary we denote by J this into isometry), one has the canonical decomposition $X^{***} = X^* \oplus X^\perp$; i.e., an element $f \in X^{***}$ admits the unique decomposition $f = f_{|X} + (f - J(f_{|X}))$. When the decomposition of $X^{***} = X^* \oplus_1 X^\perp$ is in the 1-norm; i.e, X is an M-ideal in X^{**} (see [162]) one has.

Lemma 2.10.i. *Let M be a subspace of an M-ideal in its bidual X. For every $f \in M^{\perp\perp}$ one has dist $(f, M) = $ dist (f, X).*

Proof. The inequality dist $(f, M) \geq$ dist (f, X) is clear. Let $i: M \to X$ the canonical injection and $Ji(M)$ the natural copy of M in X^{**}. Since dist $(f, M) = \| f + M \|$ in $X^{**}/Ji(M)$ one has

$$dist (f, M) \leq \| f - m \| = F(f)$$

for some norm one element $F \in Ji(M)^\perp \subset X^{***}$. The canonical decomposition of X^{***} yields $F = F_{|X} + x^\perp$ and thus if $x \in X$ is chosen such that $\| f - x \| \leq (1+\varepsilon)dist (f, X)$ then

$$F(f) = F_{|X} (f) + x^\perp(f - x) = 0 + \| x^\perp \| \, \| f - x \|$$

since $f \in M^{\perp\perp}$ and $F_{|X}$ belongs to $M^\perp$. Finally

$$\| x^\perp \| \, \| f - x \| \leq \| F \| \, (1+\varepsilon) \, dist (f, X).$$

Letting $\varepsilon \to 0$ the result is proved. $\square$

In general, one only has

Lemma 2.10.j. *Let M be a subspace of a Banach space X. For every $F \in M^{\perp\perp}$*

$$dist (F, M) \leq 2 \, dist (F, X).$$

Proof. Given $F \in M^{\perp\perp}$ select $\alpha \in J(M)^\perp$ such that $\| \alpha \| = 1$ and

$$\alpha(F) = dist(F, J(M)).$$

Since $J(M)^\perp = J(M^\perp) \oplus J(X)^\perp$, α can be decomposed as $\alpha = \beta + \gamma$ with $\beta \in J(M^\perp)$ and $\gamma \in J(X)^\perp$. Since $\alpha(F) = \gamma(F)$ and $\| \gamma \| \leq 2$ (β is a restriction of α):

$$dist(F, J(X)) \geq (1/2)\gamma(F) = (1/2)dist(F, J(M)). \quad \square$$

Now it is clear that the map $V: M^{\perp\perp}/M \to (M^{\perp\perp} + X)/X$ given by $V(f + M)$

$= f + X$ is an isomorphism, and one has the following result:

Proposition 2.10.k. *For every subspace M of X, $M^{\perp\perp} + X$ is closed in X^{**}.*

The next lemma allows us to prove some new results of stability. Let $Q: X^{**} \rightarrow X^{**}/X$ denote the natural quotient map.

Lemma 2.10.l. *Let M and N be closed subspaces of X. One has*

$$\gamma[\, Q(M^{\circ\circ}), Q(N^{\circ\circ})\,] \leq 2\gamma(M,N).$$

Proof. Given $m \in M^{\circ\circ}$ and $n \in N^{\circ\circ}$ clearly one has $\|Qm - Qn\| \leq \|m - n\|$; hence

$$dist\,(\,Qm,\, Q(N^{\circ\circ})\,) \leq dist\,(m,\, N^{\circ\circ}\,).$$

If, in addition, $\|Qm\| = 1$ then given $\varepsilon > 0$ there is some $x \in X$ such that $\|m + x\| < 1 + \varepsilon$. By the preceding proofs it is possible to find $v \in M$ such that $\|m + v\| < 2(1 + \varepsilon)$. This yields

$$
\begin{aligned}
dist[\, Qm,\, Q(N^{\circ\circ})\,] \;&=\; dist[\, Q(m+v),\, Q(N^{\circ\circ})\,] \\
&\leq\; dist(\, m+v,\, N^{\circ\circ}\,) \\
&<\; 2(1+\varepsilon)\,\gamma(M^{\circ\circ},\, N^{\circ\circ}) \\
&=\; 2(1+\varepsilon)\gamma(M,\, N),
\end{aligned}
$$

since $\gamma(M^{\circ\circ},\, N^{\circ\circ}) = \gamma(M,\, N)$ [218, IV.2.9]. Taking the supremum over the unit sphere of $Q(M^{\circ\circ})$ the result follows. □

Proposition 2.10.m. *If P is δ-stable (δ-costable), then $P^{co} = \{X: X^{**}/X \in P\}$ is $\delta/2$-stable ($\delta/2$-costable).*

Proof. It is enough to note that given a closed subspace M of X one has the isomorphisms

$$Q(M^{\circ\circ}) = (M^{\circ\circ} + X)/X = M^{\circ\circ}/(X \cap M^{\circ\circ}) = M^{**}/M$$

and then apply 2.10.1. □

Recall that a Banach space X is quasireflexive if $dim\, X^{**}/X < \infty$.

Corollary 2.10.n.
   i) The classes of quasireflexive and reflexive spaces are 1/2-stable and 1/2-costable.
   ii) The class Sep^{co} is 1/4-stable and 1/4-costable.
   iii) Let M,N be closed subspaces of X. If $\gamma(M,N) < 1/2$ then
       $dim\, N^{**}/N - dim\, M^{**}/M = dim\, (X/M)^{**}/(X/M) - dim\, (X/N)^{**}/(X/N).$

This is similar to the stability of the index of Fredholm operators [218, Thm. IV.5.17].

Perturbation of operators

The results of this section can be applied to the study of perturbations of operators with closed range. The linking is the following basic result due to Markus [256]. Recall that $\gamma(T) = inf \{ \parallel Tx \parallel \ : dist(x, \ Ker \ T)=1\}$. It is a standard result [218 Thm.IV.5.2] that $R(T)$ is closed if and only if $\gamma(T) > 0$.

Proposition 2.10.o. *Let $T, \ A \in L(X, \ Y)$. Assume $R(T)$ is closed, and let $S=T+A$.*
  i) $\gamma \ [Ker \ S, \ Ker \ T] \ \leq \ \gamma(T)^{-1} \parallel S-T \parallel$;
  ii) $\gamma \ [Im \ S, \ Im \ T] \ \leq \ \gamma(T)^{-1} \parallel S-T \parallel$.

Proof. For every $x \in X$ one has

$$\gamma(T) \ dist[\ x, \ Ker \ T \ ] \ \leq \ \parallel Tx \parallel.$$

  Thus, for $x \in N(S)$

$$dist[\ x, \ Ker \ T \ ] \ \leq \ \gamma(T)^{-1} \parallel (S-T)x \parallel;$$

and this yields *i*).

  The proof of *ii*) follows applying *i*) to T^*: since $(Im \ T)^\circ \ = \ Ker \ T^*$ then $\gamma(T)=\gamma(T^*)$ [218, IV.5.13], and $\gamma(M, \ N)=\gamma(N^\circ, \ M^\circ)$ [218, IV.2.9]. $\square$

Corollary 2.10.p. *Let $A, \ T \in L(X, \ Y)$, T with closed range and let $0 < \delta \leq 1$ such that $\parallel A \parallel < \delta\gamma(T)$. If P is δ-stable and $KerT \in P$, then $Ker(T+A) \in P$. If P is δ-costable and $Y/ImT \in P$, then $Y / \overline{Im(T+A)} \in P$.*

Proof. The first part follows from the definition of δ-stability and *i*) of the result above that gives $\gamma \ [Ker \ (T+A), \ Ker \ T] < \delta$; analogously for the second part. $\square$

  As an application of this perturbation theorem we get a refinement of the instability results in [145].

Proposition 2.10.q. *Assume P stable and such that $M \times N \in P$ when $M \in P$ and N is finite dimensional. Let $T \in L(X, \ Y)$ be an operator with closed range, and let $K \in L(X, \ Y)$ be a compact operator.*

  i) If $KerT \in P$ then $Ker(T+K) \in P$;
  ii) If $Y/ImT \in P$ then $Y / \overline{Im(T+K)} \in P$.

Proof. i) Since K is a compact operator there is a finite codimensional subspace M

of X such that $\|Ki_M\|$ is arbitrarily small [285, 3.2.3]; say, smaller than $a\gamma(T)$. Then Ti_M is a closed operator, $\gamma(Ti_M) \geq \gamma(T)$ and *Ker* $Ti_M \in P$. Hence *Ker* $(T+K)i_M \in P$ by the result above; and *Ker* $(T+K) \in P$ since it is the sum of *Ker* $(T+K)i_M$ and a finite dimensional subspace of X.

ii) It can be proved analogously, taking into account the existence of a finite dimensional subspace N of Y such that $\|q_N K\| < a\gamma(T)$, where q_N denotes the quotient map onto Y/N.   □

Questions 2.10.r. *Are they open*

1) having the r-Banach-Saks property for all $r > r_0$?
2) having the property W_r for all $r > r_0$?
3) having type r for all $r < r_0$?
4) having cotype r for all $r > r_0$?

Chapter 3

Classical Banach spaces

It is time to enter and see the monsters. In this chapter we shall study how to obtain twisted sums starting with certain Banach sequence spaces such as l_p. Two solutions to Palais problem, that of Enflo, Lindenstrauss and Pisier [116] and that of Kalton and Peck [213] shall be examined in detail. Two general problems still open to research are *how to construct quasi-linear maps between arbitrary Banach spaces?* and *how to determine if the corresponding twisted sum splits?* The techniques developed in the chapter work in certain sequence spaces. Twisted sums of function spaces shall be considered in 3.4 and 3.5.

On the side of positive $3SP$ results, the methods described in Chapter 2 shall be applied to obtain simpler proofs of the $3SP$ character of several properties related to the l_p-structure of a Banach space. In two appendices we consider two related topics: a sketch of the recent solution of the $3SP$ problem for dual spaces, and the "simplest case of failure of a $3SP$ problem": Banach spaces not isomorphic to their square.

3.1   The 3*SP* problem for Hilbert spaces

The origin of $3SP$ problems can be traced back to questions concerning the isomorphic nature of subspaces of a given space. A problem attributed (see [116] or [247] problem 4.b) to Palais is: If Y and X/Y are isomorphic to a Hilbert space, has X to be isomorphic to a Hilbert space? This is the $3SP$ problem for Hilbert spaces, answered negatively by Enflo, Lindenstrauss and Pisier [116] and later by Kalton and Peck [213]. Their results complete the solution of the $3SP$ problem for $\mathscr{L}_p$-spaces, raised in [247, Problem 4.a].

The approach of Enflo, Lindenstrauss and Pisier

The solution to the $3SP$ for l_2 spaces given by Enflo, Lindenstrauss and Pisier in [116] differs from the general schema described in 1.5 only in the form of constructing the quasi-linear (actually 0-linear) map $l_2 \to l_2$. It shall be constructed using the finite dimensional structure of l_2.

Recall from 1.6 that a homogeneous map F has been called 0-linear if for some $K > 0$ it is satisfied that whenever (x_i) is a finite set points with $\Sigma_i x_i = 0$ then

$$\| \Sigma_i F x_i \| \le K \Sigma_i \| x_i \| .$$

Consider the class B_n of all 0-linear functions $g : l_2^n \to l_2^{n^2}$ with $K \le 1$. Functions of B_n are automatically bounded: if $x = \Sigma_{1 \le i \le n} \lambda_i e_i$ then $x - \Sigma_{1 \le i \le n} \lambda_i e_i = 0$ and thus

$$\| g(x) \| \le \Sigma_{1 \le i \le n} \| \lambda_i e_i \| + \| x \| + \| \Sigma_{1 \le i \le n} g(\lambda_i e_i) \| .$$

The elements of B_n are not necessarily linear. For instance, if $n = 2$ the map

$$g(x, y) = (x, y, (x^2 + y^2)^{-1/2} x |y|, 0)$$

is 0-linear with $K \le 1/2$:

if $\Sigma_{1 \le i \le n} x_i = \Sigma_{1 \le i \le n} y_i = 0$ then

$$\| \sum_{i=1}^n g(x_i, y_i) \| = | \sum_{i=1}^n (x_i^2 + y_i^2)^{-1/2} x_i |y_i| |$$

$$\le \sum_{i=1}^n (x_i^2 + y_i^2)^{-1/2} \frac{1}{2} (x_i^2 + y_i^2)$$

$$= \frac{1}{2} \sum_{i=1}^n \| (x_i, y_i) \| .$$

From 1.6.f we know that the distance

$$d = dist(g, L_n) = \inf_{T \in L_n} \sup_{\|x\| \le 1} \| g(x) - T(x) \|$$

from g to the space of linear maps is greater than or equal to the constant K of g. The preceding calculus shows that $K \le 1/2$, so it is not enough to estimate d. However, $K \ge (2 + \sqrt{2})^{-1}$ as the following calculation shows:

$$\| g(1, 1) + g(1, -1) + g(-2, 0) \| = \sqrt{2}.$$

More generally, given $f \in B_n$ the function

$$g(x,y) = \left[f(x), f(y), \frac{x\|y\|}{\left(\|x\|^2+\|y\|^2\right)^{1/2}} \right]$$

belongs to B_{2n}:

Proof. Since the homogeneity is clear, let $\Sigma_i x_i = 0 = \Sigma_i y_i$. If we write

$$\alpha_i = \frac{\|y_i\|}{\left(\|x_i\|^2+\|y_i\|^2\right)^{1/2}}$$

then, for any scalar c one has:

$$\left\| \Sigma_i\, g(x_i, y_i) \right\|^2 = \left\| \left(\Sigma_i f(x_i),\ \Sigma_i f(y_i),\ \Sigma_i \alpha_i x_i - c\Sigma_i x_i \right) \right\|^2$$

$$= \left\| \Sigma_i f(x_i) \right\|^2 + \left\| \Sigma_i f(y_i) \right\|^2 + \left\| \Sigma_i (\alpha_i - c)x_i \right\|^2$$

$$\leq \left(\Sigma_i \|x_i\| \right)^2 + \left(\Sigma_i \|y_i\| \right)^2 + 2\left(\Sigma_i |\alpha_i - c| \|x_i\| \right)^2.$$

Put now

$$c = \frac{\Sigma_i \alpha_i \|x_i\|}{\Sigma_i \|x_i\|}.$$

By Holder's inequality

$$\left(\Sigma_i |\alpha_i - c| \|x_i\| \right)^2 \leq \left(\Sigma_i \|x_i\| \right) \left(\Sigma_i \|x_i\| (\alpha_i - c)^2 \right).$$

Observe that

$$c^{-1}\Sigma_i \alpha_i \|x_i\| = \Sigma_i \|x_i\|.$$

Therefore

$$c\Sigma_i \alpha_i \|x_i\| = c^2\Sigma_i \|x_i\|;$$

and thus

$$c^2\Sigma_i \|x_i\| - 2c\Sigma_i \alpha_i \|x_i\| = -c^2\Sigma_i \|x_i\|.$$

This yields

$$\Sigma_i \alpha_i^2 \|x_i\| + c^2 \Sigma_i \|x_i\| - 2c\Sigma_i \alpha_i \|x_i\| = \Sigma_i \alpha_i^2 \|x_i\| - c^2\Sigma_i \|x_i\|$$

where the term on the left is precisely $\Sigma_i(\alpha_i - c)^2 \|x_i\|$. Thus

$$\sum_i (\alpha_i - c)^2 \|x_i\| = \sum_i \alpha_i^2 \|x_i\| - c^2 \sum_i \|x_i\|.$$

Multiplying by $\sum_i \|x_i\|$ and taking into account that

$$c^2 \sum_{i,j} \|x_i\| \|x_j\| = \sum_{i,j} \alpha_i \alpha_j \|x_i\| \|x_j\|$$

one obtains

$$\left(\sum_i \|x_i\|\right)\left(\sum_i (\alpha_i - c)^2 \|x_i\|\right) = \frac{1}{2}\sum_i \sum_j (\alpha_i - \alpha_j)^2 \|x_i\| \|x_j\|.$$

Returning to the starting inequality, we thus obtained

$$\left(\sum_i |\alpha_i - c| \|x_i\|\right)^2 \leq \frac{1}{2}\sum_i \sum_j (\alpha_i - \alpha_j)^2 \|x_i\| \|x_j\|.$$

It is time for some device. If s and t are real numbers then the following inequality holds:

$$\left|\frac{t}{\sqrt{1+t^2}} - \frac{s}{\sqrt{1+s^2}}\right|^2 \leq 2\left(\sqrt{1+t^2}\sqrt{1+s^2} - (1+ts)\right).$$

This inequality can be proved by writing complex numbers $u = 1 + is$ and $v = 1 + it$ and computing

$$
\begin{aligned}
|t(1+t^2)^{-1/2} - s(1+s^2)^{-1/2}|^2 &\leq [Im(\,u/|u| - v/|v|\,)]^2 \\
&\leq |\,u/|u| - v/|v|\,|^2 \\
&\leq 2 - 2Re\ u\bar{v}(|u||v|)^{-1} \\
&\leq 2\,|\,(1+t^2)^{1/2}(1+s^2)^{1/2} - (1+ts)|^2\,(|u||v|)^{-1} \\
&\leq 2\left((1+t^2)^{1/2}(1+s^2)^{1/2} - (1+ts)\right).
\end{aligned}
$$

Set now $t = \|y_i\| / \|x_i\|$ and $s = \|y_j\| / \|x_j\|$. One has

$$t(1+t^2)^{-1/2} = \|y_i\| (\|x_i\|^2 + \|y_i\|^2)^{-1/2}$$

and also

$$(\alpha_i - \alpha_j)^2 \leq 2\left((1+t^2)^{1/2}(1+s^2)^{1/2} - (1+ts)\right).$$

Putting all together one gets

$$\left\| \sum_i g(x_i, y_i) \right\|^2 \leq \left(\sum_i \|x_i\| \right)^2 + \left(\sum_i \|y_i\| \right)^2 + \frac{1}{2} \sum_i \sum_j \|x_i\| \|x_j\| (\alpha_i - \alpha_j)^2$$

$$\leq \left(\sum_i \|x_i\| \right)^2 + \left(\sum_i \|y_i\| \right)^2$$

$$+ \sum_i \sum_j \left((\|x_i\|^2 + \|y_i\|^2)(\|x_j\| + \|y_j\|^2) \right)^{1/2}$$

$$- \sum_i \sum_j \|x_i\| \|x_j\| + \|y_i\| \|y_j\|$$

$$= \sum_i \|x_i\|^2 + \sum_i \|y_i\|^2 + \sum_{i \neq j} \left(\|x_i\|^2 + \|y_i\|^2)(\|x_j\|^2 + \|y_j\|^2) \right)^{1/2}$$

$$= \left[\sum_i \left(\|x_i\|^2 + \|y_i\|^2 \right)^{1/2} \right]^2$$

$$= \left(\sum_i \|(x_i, y_i)\| \right)^2 .$$

$$\square$$

Let $f: l_2^n \to l_2^{n^2}$ be a function of the class B_n and let

$$X_n = l_2^{n^2} \oplus_f l_2^n$$

be the corresponding snarked sum; i.e., the product space endowed with the norm
$[\![\ ]\!]$ induced by the closed convex hull of the points $(y, 0)$, with $\|y\| \leq 1$, and
$(f(z), z)$, with $\|z\| \leq 1$. From the first claim in 1.6.e we know that the subspace

$$Y_n = \{(y, 0): y \in Y\}$$

of X_n is isometric to $l_2^{n^2}$, and $Z_n = X_n/Y_n$ is isometric to l_2^n. Now, if $P: X_n \to Y_n$
is a continuous projection, since $P(0, x) = (Tx, 0)$ for some linear (obviously
continuous) operator $T \in L_n$ one has

$$P(f(x), x) = P(f(x), 0) + P(0, x) = (Tx, 0) + (f(x), 0),$$

from where

$$[\![P]\!] \geq \sup_{\|x\| \leq 1} [\![P(f(x), x)]\!] = \sup_{\|x\| \leq 1} [\![Tx + f(x)]\!] \geq dist(f, L_n).$$

Thus, the required construction shall be obtained making the nonlinearity of f
increase as $n \to \infty$. This is the content of the following lemma.

Lemma. *There is a constant $C > 0$ such that for every n there is a function g_n in B_n
such that $dist(g_n, L_n) \geq C(\log n)^{1/2}$.*

Proof. Let $D_n = \sup\{\,dist\,(f, L_n): f \in B_n\,\}$. The key point is to prove that, for some $\delta > 0$,

$$D_{2n}^2 \geq D_n^2 + \delta,$$

since, being this true, a simple induction shows

$$D_{2^{k+1}}^2 = D_{2^k 2}^2 \geq D_{2^k}^2 + \delta \geq C(\log 2^k)^{1/2} + \delta = C\sqrt{k} + \delta$$

$$\geq C\sqrt{k+1} = C(\log 2^{k+1})^{1/2}$$

(the existence of C independent of k is consequence of $\lim_{k\to\infty}\sqrt{k+1} - \sqrt{k} = 1$). From here, the result holds for all integers replacing C by, say, $C/2$: if m is given and k is the greatest integer for which $2^k \leq m$ then

$$D_m^2 \geq D_{2^k}^2 \geq C(\log 2^k)^{1/2} = C\sqrt{k} = \frac{C}{\sqrt{2}}\sqrt{2k} \geq \frac{C}{\sqrt{2}}(\log m)^{1/2}.$$

Therefore, it remains to prove that $\delta = (2+\sqrt{2})^{-2}$ works.

Claim: $D_{2n}^2 \geq D_n^2 + (2+\sqrt{2})^{-2}$.

Proof. There is no loss of generality assuming that for some $f \in B_n$ the equality $D_n(f) = D_n$ holds; this avoids to take care of several ε. Let $g \in B_{2n}$ defined as before by

$$g(x,y) = \left[f(x), f(y), \frac{x\|y\|}{(\|x\|^2 + \|y\|^2)^{1/2}} \right].$$

The plan is to prove that

$$D_{2n}^2(g) \geq D_n^2(f) + \delta.$$

Let $T: l_2^{2n} \to l_2$ be linear. We write T in the form

$$T(x,0) = (U_1 x, U_2 x, U_3 x)$$

$$T(0,y) = (V_1 y, V_2 y, V_3 y).$$

By definition of $D_n(f)$, and with the same assumption as before, there are norm-one points x_0 and y_0 in l_2^n so that

$$\|U_1 x_0 - f(x_0)\| \geq D_n(f)$$

$$\|V_2 y_0 - f(x_0)\| \geq D_n(f).$$

Evaluating g at the point $(x_0, 0) \in l_2^{2n}$ one gets

$$D_{2n}^2 \geq D_{2n}^2(g) \geq \| U_1 x_0 - f(x_0) \|^2 + \| U_3 x_0 \|^2 \geq D_n(f)^2 + \| U_3 x_0 \|^2.$$

Consider next the points $\dfrac{1}{\sqrt{2}}(x_0, \pm y_0) \in l_2^{2n}$. The definition of g and the 0-linearity of f give

$$g\left[\frac{1}{\sqrt{2}}x_0, \frac{1}{\sqrt{2}}(\pm y_0)\right] - T\left[\frac{1}{\sqrt{2}}x_0, \frac{1}{\sqrt{2}}(\pm y_0)\right] =$$

$$= \frac{1}{\sqrt{2}}\left[f(x_0) - U_1 x_0, \ -U_2 x_0, \ \frac{1}{\sqrt{2}}x_0 - U_3 x_0\right]$$

$$\pm \frac{1}{\sqrt{2}}(V_1 y_0, \ V_2 y_0 - f(y_0), \ V_3 y_0).$$

Since $max\{ \| z+w \|^2, \| z-w \|^2\} \geq \| z \|^2 + \| w \|^2$, one has

$$D_{2n}^2 \geq D_{2n}^2(g) \geq \frac{1}{2}(\ \| f(x_0) - U_1 x_0 \|^2 + \| U_2 x_0 \|^2 + \| \frac{1}{\sqrt{2}}x_0 - U_3 x_0 \|^2$$

$$+ \| V_1 y_0 \|^2 + \| V_2 y_0 - f(y_0) \|^2 + \| V_3 y_0 \|^2 \ )$$

$$\geq D_n^2 + \| \frac{1}{2}x_0 - \frac{1}{\sqrt{2}}U_3 x_0 \|^2.$$

Thus, one arrives to

$$D_{2n}^2 \geq D_n^2 + \| U_3 x_0 \|^2$$
$$D_{2n}^2 \geq D_n^2 + \| \frac{1}{2}x_0 - U_3 \frac{1}{\sqrt{2}}x_0 \|.$$

Now, $1 = \| x_0 \|$

$$\leq \| U_3 \sqrt{2} x_0 \| + \| x_0 - U_3 \sqrt{2} x_0 \|$$

$$= \sqrt{2} \| U_3 x_0 \| + 2 \| (1/2)x_0 - U_3 (1/\sqrt{2})x_0 \| ;$$

but if a and b are such that $1 \leq \sqrt{2}a + 2b$ then either a or b is greater than or equal to $(2+\sqrt{2})^{-1}$.   □

To finish with the construction, choose a sequence $g_n \in B_n$ of functions such that $dist(g_n, L_n)$ goes to infinity, let X_n be the twisted sum

$$X_n = l_2^{n^2} \oplus_{g_n} l_2^n$$

and let $Y_n = l_2^{n^2}$. The space

$$X = \left(\sum_n X_n\right)_2$$

contains the subspace

$$Y = \left(\sum_n Y_n\right)_2$$

that is isometric to l_2, and such that

$$Z = \left(\sum_n Z_n\right)_2$$

is also isometric to l_2. The space X is not isomorphic to l_2 since Y is not complemented in X.   □

The approach of Kalton and Peck

Consider E a normed space of sequences with the following properties:

 i)   the space φ of finite sequences is dense in E;
 ii) $\|e_n\| = 1$;
 iii)   if $s \in l_\infty$ and $x \in E$ then $\|sx\| \leq \|s\|_\infty \|x\|_E$;
 iv) $\|x\|_\infty \leq \|x\|$

A method to obtain quasi-linear maps $E \to E$ was given by Kalton and Peck. They start with a lipschitz map $\phi\colon \mathbb{R} \to \mathbb{R}$ such that $\phi(t)=0$ for $t\leq 0$, pass to a quasi-additive (to be defined) function in φ (the space of finite sequences) and then to a quasi-linear map in the completion of φ, i.e., in E.

A *quasi-additive* map f is a map (not necessarily homogenous) that is *symmetric*, in the sense that $f(-x) = -f(x)$, and *continuous at* 0, in the sense that $lim_{t \to 0} f(tx)$ $= 0$, and such that for some $K>0$ and all points x, y, verifies

$$\| f(x+y) - f(x) - f(y) \| \leq K(\|x\| + \|y\|)$$

Step 1. A lipschitz map $\phi\colon \mathbb{R} \to \mathbb{R}$ with lipschitz constant L_ϕ induces a quasi-additive function $f\colon \varphi \to E$ of constant $L_\phi \log 2$ by the formula

$$f(x)(n) = x(n)\phi(-\log |x(n)|)$$

where $f(x)(n)=0$ if $x(n)=0$.

Proof. One only needs the pointwise estimate

$$|f(x)(k) + f(y)(k) - f(x+y)(k)| \leq L_\phi |x(k)+y(k)| \, log \, 2.$$

and the lattice structure of E □

Step 2. A quasi-additive function $f: \varphi \to E$ induces a quasi-linear map $F: \varphi \to E$ by the formula

$$F(x) = \|x\| \, f(x/\|x\|).$$

Proof. Observe that if $f: \mathbb{R} \to E$ is a quasi-additive map then it is bounded on compact sets, i.e., if $a \leq b$ then

$$sup_{t \in [a, b]} \|f(t)\| < \infty:$$

since otherwise there must be some sequence (t_n) in $[a, b]$ convergent to some point t and such that $\|f(t_n)\| > n$. We shall show that $f(t)$ does not exist: let K be the constant of f; since

$$\|f(t) - f(t-t_n) - f(t_n)\| \leq K(|t-t_n| + |t_n|) \leq K(2b-a)$$

one has

$$\|f(t_n)\| \leq K(2b-a) + \|f(t)\| + \|f(t-t_n)\|.$$

Taking limits when $n \to \infty$, the quasi-additivity of f makes $\|f(t-t_n)\| \to 0$ and this leads to $2n - K(2b-a) \leq \|f(t)\|$ for large n: an absurd.

This simple argument does not show, however, that if $f(a)=f(b)=0$ the bound for f on a given interval $[a,b]$ only depends on its constant K and on the space E. Let us show this. Without loss of generality we shall work on $[0,1]$. That $f(0)=0$ is consequence of the quasi-additivity of f, and we assume that $f(1)=0$. With this, one can inductively prove

$$\|f(2^{-(n+1)})\| \leq K2^{(n+1)} + 2^{-1} \|f(2^{-n})\|.$$

Now the estimate in 1.6 can be used to obtain

$$\|f(\textstyle\sum_n 2^{-n})\| \leq K\sum_n n2^{-n} + \|\sum_n f(2^{-n})\|.$$

where the second summand can be inductively estimated by

$$\|\textstyle\sum_n f(2^{-n})\| \leq \sum_n \|f(2^{-n})\| \leq \sum_n K2^{-n} + \frac{1}{2}\sum_n \|f(2^{-n+1})\|$$

$$\leq \textstyle\sum_n K2^{-n} + \frac{1}{2}\sum_n K2^{-n+1} + \frac{1}{4}\sum_n \|f(2^{-n+2})\|$$

$$\cdots$$

$$\leq 2K$$

To finish, write an element $t \in [0, 1]$ as $t = \sum_n \dfrac{\delta_n}{2^n}$ where δ_n is 0 or 1.

Returning to the proof, observe that the quasi-additive map $t \to f(t) - tf(1)$ takes the value 0 at 0 and 1. Now, if $f\colon \varphi \to E$ is a quasi-additive map, composition with the linear map $t \to tx/\|x\|$ gives a quasi-additive map $\mathbb{R} \to E$ with the same constant as $f(t) - tf(1)$ and thus bounded on $[0, 1]$ by some bound M that only depends on K and E and not on the point x:

$$sup_{t \in [0,1]} \|f(tx/\|x\|) - tf(x/\|x\|)\| \leq M.$$

Thus, since $\|x\| + \|y\| \leq 1$ implies $\|x+y\| \leq 1$ and $\|x\| \leq 1$, one has

$\|F(x+y) - f(x+y)\| \leq M,$

$\|F(x) - f(x)\| \leq M$, and

$\|f(x+y) - f(x) - f(y)\| \leq K$ (by the quasi-additivity of f).

All this yields

$\|F(x+y) - F(x) - F(y)\| \leq$

$\qquad \leq \|F(x+y) - f(x+y)\| + \|F(x) - f(x)\| + \|F(y) - f(y)\| + K$

$\qquad \leq 3M + K.$ $\qquad\qquad\qquad\qquad\qquad\qquad\qquad\qquad\qquad\qquad$ $\square$

Quasi-linear maps can be extended from dense subspaces:

Step 3. *A quasi-linear map* $D\colon Z_0 \to Y$ *defined on a dense subspace of* Z *can be extended to a quasi-linear map* $H\colon Z \to Y$.

Proof. The quasi-linear map D allows one to construct the twisted sum $Y \oplus_D Z_0$, which can be completed as a quasi-normed space X. It is not hard to verify that X is a twisted sum of Y and Z. The quasi-Banach space X is thus equivalent to a twisted sum $Y \oplus_F Z$ for some quasi-linear map $F\colon Z \to Y$. Therefore, for some linear continuous map T one has the commutative diagram:

$$0 \to Y \to X \to Z \to 0$$
$$\| \qquad \uparrow T \qquad \|$$
$$0 \to Y \to Y \oplus_F Z \to Z \to 0$$

which means that for some linear map $A\colon Z \to Y$ the operator T has the form

$$T(y, z) = (y + Az, z);$$

therefore, for $z \in Z_0$

$$\| Az+Fz-Dz \| \leq \| Az+Fz-Dz \| + \| z \|$$
$$= \| (Az+Fz, z) \|$$
$$= \| T(Fz, z) \|$$
$$\leq \| T \| \| (Fz, z) \|$$
$$= \| T \| \| z \| .$$

Thus, defining $H(z)=D(z)$ when $z \in Z_0$ and $H(z)=F(z)+A(z)$ otherwise, the map H is a quasi-linear extension of D to Z. $\quad\square$

Now, the most delicate point appears: as we already know, the twisted sums $Y \oplus_F Z$ and $Y \oplus_G Z$ are equivalent if and only if F and G are related by some linear (not necessarily continuous) map A in the form: $F-G-A$ is bounded on the unit ball. When F and G come induced by Lipschitz maps ϕ and ψ as above, how is the equivalence expressed in terms of the Lipschitz maps?

Assume that E satisfies an

Additional assumption

[♥] *No subsequence of the canonical basis* (e_n) *is equivalent to the canonical basis of* c_0

then one has that if E_ϕ is the twisted sum of E with E with respect to the quasi-linear map induced by ϕ in the form explained so far then

Proposition 3.1.a. *The twisted sums* E_ϕ *and* E_ψ *are equivalent if and only if*

$$sup_{t \in \mathbf{R}} \, |\phi(t) - \psi(t)| \, < \, \infty.$$

Proof. It is easy to see that, by the definition of F and G,

$$\| F(x)-G(x) \| \leq sup_{t \in \mathbf{R}} |\phi(t) - \psi(t)| \, \| x \|$$

and thus E_ϕ and E_ψ are equivalent if $\phi-\psi$ is bounded (choice $A=0$) .

To obtain the converse implication, assume that E_ϕ and E_ψ are equivalent, i.e., for some linear map $A: E \to E$, the difference $F-G-A$ is bounded on the unit ball by some constant, say M. Then since $F(e_n)=G(e_n)=0$, $\| Ae_n \|_\infty \leq \| Ae_n \|_E \leq M$. Since bounded sequences in l_∞ admit pointwise convergent subsequences, it is possible to obtain increasing sequences of integers $(n(k))_k$ and $(m(k))_k$ such that $(A(e_{n(k)}-e_{m(k)}))$ is pointwise convergent to 0. Taking into account that (e_n) is a

Schauder basis for E, a gliding hump technique allows one to select a subsequence (f_n) of $(e_{n(k)} - e_{m(k)})_k$ and an increasing sequence (p_n) of indices such that

(1) $\left\| Af_n - \sum_{k=p_{n-1}+1}^{p_n} \left(Af_n(k) \right) e_k \right\| \leq 2^{-n}$

(2) $support \, f_n = \{ q_n, r_n \} \subset \{ p_{n-1}+1, \ldots, p_n \}$.

Define now an auxiliary (linear) map B: $span \, \{f_n\}_{n \in \mathbb{N}} \to E$ by

$$Bf_n = (Af_n)(q_n) e_{q_n} + (Af_n)(r_n) e_{r_n}.$$

One has:

$$\| Bf_n \|_\infty \leq \| Bf_n \|_E = \| \chi_{\{q_n, r_n\}} Af_n \|_E$$

$$\leq \| Af_n \|_E$$

$$\leq \left(\| Ae_{q_n} \|_E + \| Ae_{r_n} \|_E \right)$$

$$\leq 2M \, ;$$

and, since the support of Bf_n coincides with the support of f_n,

$$|Bf_n| \leq M|f_n|.$$

Let $x = \sum_{1 \leq n \leq N} t_n f_n \in span\{f_n\}$; then clearly

$$|t_n| \leq \| x \|_\infty \leq \| x \|_E$$

and thus

$$\left\| A \left(\sum_{n=1}^N t_n f_n \right) - \sum_{n=1}^N t_n \, Bf_n \right\|_E$$

$$= \left\| \sum_{n=1}^N t_n Af_n - \sum_{n=1}^N t_n \sum_{k=p_{n-1}+1}^{p_n} \left(Af_n(k) \right) e_k \right\|_E$$

$$\leq \sum_{n=1}^N |t_n| \left\| Af_n - \sum_{p_{n-1}+1}^{p_n} \left(Af_n(k) \right) e_k \right\|_E$$

$$\leq \left(\sum_{n=1}^N 2^{-n} \right) \max_{n \leq N} |t_n|$$

$$\leq \| x \|_E.$$

Therefore

$$\| F(x) - G(x) - \sum_{n=1}^{N} t_n B f_n \|_E \leq \| F(x) - G(x) - Ax + Ax - \sum_{n=1}^{N} t_n B f_n \|_E$$

$$\leq (M+1) \| x \|.$$

Hence, if $s_n = f_1 + \ldots + f_n$ then

$$\| F(s_n) - G(s_n) - \sum_{n=1}^{N} B f_n \| \leq (M+1) \| x \|.$$

Now, observe that if $1 \leq k \leq n$ then

$$F(s_n)(k) = \| s_n \| f(\frac{s_n}{\| s_n \|})(k) = \phi(-\log \frac{1}{\| s_n \|}) = \phi(\log \| s_n \|),$$

and thus

$$F(s_n) - G(s_n) = (\phi(\log \| s_n \|) - \psi(\log \| s_n \|)) s_n.$$

Since $\| |x| \|_E = \| x \|_E$ one has

$$\| F(s_n) - G(s_n) - \sum_{n=1}^{N} B f_n \| =$$

$$= \| |\phi(\log \| s_n \|) - \psi(\log \| s_n \|) s_n - \sum_{n=1}^{N} B f_n | \|_E$$

$$\geq \| |\phi(\log \| s_n \|) - \psi(\log \| s_n \|)| |s_n| - \sum_{n=1}^{N} |B f_n| \|_E,$$

and this yields

$$\| |\phi(\log \| s_n \|) - \psi(\log \| s_n \|)| |s_n| - 2M |s_n| \|$$

$$\leq \| |\phi(\log \| s_n \|) - \psi(\log \| s_n \|)| |s_n| - \sum_{n=1}^{N} |B f_n| \|_E$$

$$\leq (M+1) \| s_n \|_E,$$

which gives

$$(|\phi(\log \| s_n \|) - \psi(\log \| s_n \|)| - 2M) |s_n| \leq (M+1) |s_n|.$$

Now the additional hypothesis [♥] goes at work: if no subsequence of (e_n) is equivalent to the canonical basis of c_0 then $sup \| s_n \| = +\infty$. Moreover,

$$\| s_{n+1} \| \le \| s_n \| + \| f_{n+1} \| \le \| s_{n+1} \| + 2$$

and thus, for large n,

$$\log \| s_{n+1} \| \le \log \| s_n \| + 1$$

and so

$$\log \| s_{n+1} \| - \log \| s_n \| \le 1.$$

If t is a positive real number and s_n is chosen so that

$$\log \| s_n \| \le t < \log \| s_{n+1} \|$$

then

$$|\phi(t) - \psi(t)| \le | \phi(\log \| s_{n+1} \|) - \phi(\log \| s_n \|) | + | \phi(\log \| s_n \|) - \psi(\log \| s_n \|) |$$

$$\le L_\phi(\log \| s_{n+1} \| - \log \| s_n \|) + |\phi(\log \| s_n \|) - \psi(\log \| s_n \|)|$$

$$\le 1 + |\phi(\log \| s_n \|) - \psi(\log \| s_n \|)| .$$

In other words,

$\phi - \psi$ *is bounded*   if and only if $(\phi(\log \| s_n \|) - \psi(\log \| s_n \|))_n$ *is bounded.*

But the boundedness of $(\phi(\log \| s_n \|) - \psi(\log \| s_n \|))_n$ can be obtained as follows:

If $|\phi(\log \| s_n \|) - \psi(\log \| s_n \|)| \le 2M$ then there is nothing to prove. If not, the estimate

$$(|\phi(\log \| s_n \|) - \psi(\log \| s_n \|)| - 2M) |s_n| \le (M+1)|s_n|$$

yields $|\phi(\log \| s_n \|) - \psi(\log \| s_n \|)| \le 2M + M + 1 .$ $\square$

Therefore, a non trivial twisted sum of Hilbert spaces can be obtained starting with *any* unbounded Lipschitz real function. Since Hilbert spaces are B-convex, the twisted sum can be renormed to be a Banach space. Since the short sequence does not split, the twisted sum is not trivial and, a fortiori, it is not a Hilbert space.

Weak Hilbert spaces

Refinements of the notion of Hilbert space have been introduced by Pisier [294] and named weak Hilbert and asymptotically Hilbert spaces. In [257], Mascioni observes that an Orlicz sequence space l_M is weak Hilbert if and only if it is isomorphic to

l_2 ; i.e., if M is an Orlicz function equivalent to $f(x)=x^2$. Since the space Z_2 contains an Orlicz sequence subspace which is not a Hilbert space: that spanned by $(0,e_n)_n$ (see the next paragraph), it follows that Z_2 is not a *weak Hilbert space*.

Theorem 3.1.b. *To be a weak Hilbert space is not a 3SP.*

Twisted Hilbert spaces. A question that seems to be out of reach emerged from several conversations with D. Yost [365]:

Question. *Is the property of being a twisted sum of Hilbert spaces a 3SP property?*

3.2   3*SP* problems related to l_p spaces

The approach of Kalton and Peck (and, of course, that of Enflo, Lindenstrauss and Pisier) works for any other l_p space $1<p<\infty$.

Let $\phi: \mathbb{R} \rightarrow \mathbb{R}$ be the lipschitz function defined by $\phi(t)=t$ when $t>0$ and $\phi(t)=0$ otherwise. Following 3.1, ϕ induces the quasi-linear map $F: l_p \rightarrow l_p$ given by $F(x) = \| x \| f(x/ \| x \|)$ where $f(x)(n)=x(n) \phi(-log \ |x(n)|)$. This means that

$$F\left(\sum_{i=1}^{N} e_i\right) = (\log \sqrt[p]{N}) \sum_{i=1}^{N} e_i \ .$$

Let us call Z_p the corresponding twisted sum. That the Kalton-Peck space Z_p is not isomorphic to l_p can be seen looking at the sequence $y_n=(0, e_n)_n$, which satisfies the estimate

$$\| (0, \sum^N e_j) \| = \| \sum^N e_j \|_{l_p} + \| F\left(\sum^N e_j\right) \|_{l_p} = (1+\sqrt[p]{N}) \log \sqrt[p]{N} \ .$$

Note that this estimate already implies that the sequence $(0, e_n)_n$ is weakly null.

Kalton and Peck even proved [213] that Z_p admits no complemented subspace spanned by an unconditional basic sequence. Thus, the following result has been proved:

Theorem 3.2.a. *To be isomorphic to a normable sequence space E satisfying the properties i)-iv) and such that no subsequence of the canonical basis is equivalent to the usual basis of c_0 is not a 3SP property.*

In particular

Theorem 3.2.b. *To be isomorphic to l_p is not a 3SP property for $1<p<\infty$.*

A singular instance to which this result applies is Tsirelson spaces (see [53]). It would be interesting to know the structure of twisted sums of Tsirelson-like spaces.

If one asks what happens for c_0 (when the method does not provide a way to decide if the twisted sum splits) or l_1 (for which we already know that non-locally convex twisted sums of l_1 exist), the solution of the $3SP$ problem follows from universal properties:

Theorem 3.2.c. *To be isomorphic to l_1, l_∞ or c_0 is a 3SP property.*

Proof.

case l_1: the space l_1 is projective and thus the kernel of any quotient map $X \rightarrow l_1$ is a complemented subspace of X.

case l_∞: the space l_∞ is injective and thus complemented wherever it is found.

case c_0: if Y and X/Y are isomorphic to c_0, X must be separable (a $3SP$ property) and Y must be complemented (Sobczyk's theorem, see [354]).$\square$

Not containing l_p ; and l_p saturated spaces.

Theorem 3.2.d. *For $1 \leq p < \infty$ the property of not having subspaces isomorphic to l_p, and the property of being l_p-saturated are 3SP properties. Analogously, for $1 < p < \infty$, the properties of not having quotients isomorphic to l_p, and the property of having all quotients l_p-saturated are 3SP properties.*

Proof. All the cases altogether follow from the method of incomparability 2.8: since the space l_p is minimal, taking $\Lambda = \{spaces\ isomorphic\ to\ l_p\}$ one obtains that $\Lambda(i)$ and $\Lambda(ii)$ must be $3SP$ space ideals. Observe that l_p is also co-minimal for $1 < p < \infty$. The case $p = 1$ could also follow from method 2.4.b. and Rosenthal's l_1 theorem.

Theorem 3.2.e. *The property of not having subspaces or quotients isomorphic to c_0 and the property of being c_0-saturated, or having all quotients c_0-saturated are 3SP properties.*

Proof. It is a well-known result of Bessaga and Pelczynski that a Banach space X does not contain c_0 if and only if weakly 1-summable sequences are norm convergent. An elementary form of lifting results proves that this is a $3SP$ property. See 6.1 for details. The space c_0 is minimal and co-minimal, and thus the results can be obtained by incomparability. $\square$

These results have been obtained in the context of Fréchet spaces by Díaz [83].

Theorem 3.2.f. *Not containing copies of l_∞ is a 3SP property.*

Proof. Let X be a Banach space containing an isomorphic copy of l_∞ through some isomorphism $J: l_\infty \to X$. Let Y be a subspace and $q: X \to X/Y$ the corresponding quotient map. If X/Y does not contain l_∞ then qJ is weakly compact by a result of Rosenthal (see [248, 2.f.4]):

If $T: l_\infty \to E$ is not weakly compact then it is an isomorphism when restricted to a certain subspace F of l_∞ isomorphic to l_∞.

If qJ is weakly compact then it is completely continuous (by the Dunford-Pettis property of l_∞). Since $lim \ \|qJe_n\| = 0$ we can assume that $\|qJe_n\| \le K2^{-(n+1)}$, where K denotes the basic constant of $(Je_n)_n$ in X; if we choose elements $(y_n)_n$ in Y so that $\|Je_n - y_n\| \le K2^{-(n+1)}$, the sequence $(y_n)_n$ is basic in Y and spans c_0.

Claim: A weakly compact operator $l_\infty \to Z$ is weak*-to-weak continuous when restricted to $l_\infty(M)$ for certain infinite subset $M \subset N$.

This is a re-writing of a result of Drewnowski [101]. Set $l_\infty = l_\infty(M)$ from now on. For all $A \subset M$

$$qJ(\chi_A) = weak \ lim \sum_{n \in A} qJ(e_n) \ .$$

Let us define an operator $T: l_\infty \to Y$ by means of

$$T(\chi_A) = J(\chi_A) + weak \ lim \sum_{n \in A} (y_n - Je_n)$$

Since $\|T\| \le 1 + 1/2$, T can be extended to the whole l_∞. It takes values on Y because $qT = 0$. The operator T is an isomorphism since

$$\|T\chi_A\| = \|T\chi_A - J\chi_A + J\chi_A\| \ge \|J\chi_A\| - \|(T-J)(\chi_A)\| \ge 1 - \frac{1}{2} = \frac{1}{2} \ .$$

The proof here presented is a version of Drewnowski and Roberts unpublished one [104]. An immediate consequence is the following result of Lindenstrauss and Rosenthal [248, 2.f.13]: *Let Y be a subspace of l_∞ such that l_∞/Y is reflexive. Then Y has a subspace isomorphic to l_∞.*

To be a subspace of c_0 or l_∞. Johnson and Zippin [196] observe

Theorem 3.2.g. *To be a subspace of c_0 is a 3SP property*

Proof. If Y and X/Y are subspaces of c_0, they are separable and so is X. Since c_0 is complemented in any separable space containing it, the isomorphism into $Y \to c_0$ extends to an operator $T: X \to c_0$. If $j: X/Y \to c_0$ denotes the isomorphism into and

$q\colon X \to X/Y$ is the quotient map, it is easy to verify that $T \oplus jq\colon X \to c_0 \oplus c_0$ is an isomorphism into. $\square$

The same proof without separability restrictions provides the first part of

Theorem 3.2.h. *To be a subspace of l_∞ is a 3SP property. To be a quotient of l_∞ is not a 3SP property.*

while the second part is proved in 5.21.c, where it is shown that Z_2 is not a quotient of l_∞.

Some interesting questions can be formulated:

Questions. *Is "to be a subspace of l_1" a 3SP property? Is "not having l_∞ as a quotient" a 3SP property?* For this last question see after 4.19.e. Also, observe that *"having l_∞ as a quotient" is equivalent to "having $l_1(\omega_1)$ as a subspace".* So, one is asking whether the incomparability methods in chapter 2 extend to the nonseparable case.

Local properties. Rakov passes in [302] to study the corresponding super-properties, and observes that it follows from Krivine's theorem [235] that

$$\text{super-}(not \ containing \ l_p) \ = \ not \ containing \ l_p^n \ uniformly,$$

in the sense that for every sequence (X_n) of n-dimensional subspaces of X one has:

$$\lim d(X_n, \ l_p^n) \ = \ \infty.$$

It follows that all of them are *3SP properties*.

Theorem 3.2.i. *Not containing l_p^n uniformly is a 3SP property for $1 \leq p \leq \infty$.*

3.2.j. B-convexity (K-convexity). For $p=1$, it was proved by Pisier in [289] that *not containing l_1^n-uniformly* is equivalent to *K-convexity* as defined in [258]. Giesy defined in [130] *B-convexity* as follows: a Banach space is said to be *B-(k,ε) convex* if $k^{-1}b_k \leq (1-\varepsilon)$, i.e., if given elements $x_1,\ldots,x_k$ in the unit ball there is a choice of signs so that $\| \pm x_1 \pm x_2 \ldots \pm x_k \| \leq k(1-\varepsilon)$. Therefore,

$$\text{super-}\{not \ containing \ l_1\} \ = \ B\text{-}convexity \ = \ K\text{-}convexity,$$

and thus

Theorem 3.2.k. *B-convexity is a 3SP property.*

This result was proved by Giesy [130, Thm. 9] in the context of Banach spaces and extended by Kalton [203] to locally bounded spaces (see 1.5.i).

3.2.1 C-convexity. For $p = +\infty$,

super-{*not containing* c_0} = *not containing* l_∞^n-*uniformly* = *C-convexity*

of Rakov [301] i.e.: $lim_{n \to \infty} inf d(X_n, l_\infty^n) = \infty$. Thus:

Theorem 3.2.m. *C-convexity is a 3SP property.*

Since *super*-{*not containing* c_0} = *finite cotype*, this argument appears in [90] to prove that finite cotype is a 3SP property: Let U be a nontrivial ultrafilter. If in the short exact sequence $0 \to Y \to X \to Z \to 0$ both Y and Z have finite cotype then their ultrapowers Y_U and X_U do not contain c_0, and since the sequence $0 \to Y_U \to X_U \to Z_U \to 0$ is exact, X_U does not contain c_0 and X has finite cotype. □

Question. *Is having finite cotype an open property?* An affirmative answer would provide an alternative proof for this last result.

Kislyakov [221] defines the RM_p property (after Rosenthal and Maurey), $2 < p \leq \infty$, of a space X when for all $C > 0$ and all $n \in \mathbb{N}$ there are operators $T_1 \in L(l_p^n, X)$, $T_2 \in L(X, l_\infty^n)$ in such a way that $T_2 T_1 = id : l_p^n \to l_\infty^n$ and $\| T_1 \| \ \| T_2 \| \leq C$. This could be understood as a kind of "containing l_p^n uniformly complemented." In fact, for $p = \infty$ that is exactly what it is. He announces that *Not having the RM_p is a 3SP property.*

3.3   The 3SP problem for $\mathcal{L}_p$-spaces

The spaces Z_p constructed so far provide counterexamples for some other weaker forms of "being an l_p space."

Let $1 \leq p \leq \infty$. A Banach space X is said to be an $\mathcal{L}_p$-*space* if for every finite dimensional subspace F of X there is a finite dimensional subspace G of X such that $F \subset G$ and the *Banach-Mazur distance* $d(G, l_p^n) \leq K$, where $n = dim \ G$ and K is some fixed constant depending only of X. Recall that the Banach-Mazur distance between two isomorphic spaces X and Y is defined by

$$d(X, Y) = log \ inf \ \{ \| T \| \ \| T^1 \| : T: X \to Y \text{ is an isomorphism}\}.$$

The properties of $\mathcal{L}_p$-spaces we shall need can be seen in [247] and are:

(1) *A Banach space X is a $\mathcal{L}_p$-space ($1 \leq p \leq \infty$) if and only if $X*$ is a $\mathcal{L}_{p*}$-space*

(2) *Any complemented subspace of a $\mathcal{L}_p$-space is a $\mathcal{L}_p$-space.*

(3) *For some constant K depending only on the $\mathcal{L}_p$-space X and for any finite*

dimensional subspace F of X there is a finite dimensional subspace E of X containing E and complemented with projection of norm $\leq K$.

Therefore, taking into account that if one replaces 2 by any other $1<p<\infty$ the arguments that show that projections onto Y_n have norm greater than or equal to $C(logn)^{1/p}$ remain valid with the estimate $C(logn)^{1/p}$, one has

Theorem 3.3.a. *To be a $\mathscr{L}_p$-space is not a 3SP property for $1<p<c$.*

The cases $p=1$, ∞ were solved by Lindenstrauss and Rosenthal [247]:

Theorem 3.3.b. *To be an $\mathscr{L}_p$-space is a 3SP property for $p=1$, ∞.*

Proof. Let $0 \to Y \to X \to Z \to 0$ be an exact sequence in which Y and Z are $\mathscr{L}_1$-spaces. Since injectivity is a 3SP property, *dual injective* is also a 3SP. But it turns out that $\mathscr{L}_1$-spaces are dual injective: if Y is an $\mathscr{L}_1$-space there is a factorization $Y \to L_1(\mu) \to Y^{**}$ of the canonical injection of Y into Y^{**} through some space $L_1(\mu)$. The dual factorization yields that Y^* is a complemented subspace of $L_1(\mu)^*$ and thus injective. This implies that X^* is a $\mathscr{L}_\infty$-space and thus X is a $\mathscr{L}_1$-space.

The proof for $\mathscr{L}_\infty$-spaces follows by duality.   □

A different approach was made by Jarchow [184]: the identifications

$$\mathscr{L}_1\text{-spaces} = space\ ([\mathscr{F}_1 \circ K] \circ K^{-1}),$$

$$\mathscr{L}_\infty\text{-spaces} = space\ (K^{-1} \circ [K \circ \mathscr{F}_\infty])$$

and the method 2.7 prove that they are 3SP ideals.

3.4   3SP problems for L_p-spaces

Regarding function spaces, one has:

Theorem 3.4.a. *To be isomorphic to $L_p(0, 1)$ is not a 3SP property for $1<p<\infty$.*

Proof. Taking into account that $L_p = l_p \oplus L_p = L_p \oplus L_p$, observe the exact sequence

$$0 \to l_p \oplus L_p \to Z_p \oplus L_p \oplus L_p \to l_p \oplus L_p \to 0.$$ □

Theorem 3.4.b. *To be isomorphic to $L_1(\mu)$ is a 3SP property.*

Proof. Let $0 \to L_1(\mu) \to X \to L_1(\mu) \to 0$ be an exact sequence. Since $\mathscr{L}_1$-spaces are dual injective, the dual sequence splits and so does the bidual sequence. On the other hand, an $L_1(\mu)$-space is complemented in its bidual, so in X^{**} and then in X.

Thus the sequence $0 \to L_1(\mu) \to X \to L_1(\mu) \to 0$ splits and X is isomorphic to $L_1(\mu) \oplus L_1(\mu)$ which, by general representation theorems is isomorphic to $L_1(\mu)$ □

Problem. *Is the property of not containing subspaces isomorphic to $L_p(0, 1)$, $1 < p < \infty; p \neq 2$, a 3SP property?*

The 3SP problem for *not containing L_1* raised in [246] was answered negatively by Talagrand [342].

Theorem 3.4.c. *Not containing L_1 is not a 3SP property.*

Actually, Talagrand proved:

Theorem 3.4.d. *There exist two subspaces (isomorphic to l_1-sums of subspaces isomorphic to l_1) of L_1, X_1 and X_2, such that L_1/X_1 and L_1/X_2 do not contain L_1, and such that $L_1/X_1 \times L_1/X_2$ contains L_1.*

The proof of Talagrand is a masterpiece that combines an abstract frame with very delicate hard-analysis constructions from which we can only give a

Faint sketch of proof:

First portion of the proof: An abstract setting. Let λ denote the measure of Lebesgue and Σ the σ-algebra of measurable sets. Let $\Sigma' \subset \Sigma$ be an atomless subalgebra and $A \in \Sigma'$ with $\lambda(A) > 0$; $L_{1,0}(A, \Sigma')$ denotes the space of Σ'–medible functions that vanish outside A and such that $\int f \, d\lambda = 0$. One of the main technical devices used in [342] is a result of Rosenthal [314, Thm. 1.1]:

Proposition 3.4.e. *Let $T: L_1 \to E$ be an operator that fixes a copy of L_1. There exist $\delta > 0$, $\Sigma' \subset \Sigma$ atomless, $A \in \Sigma'$ with $\lambda(A) > 0$ such that, for all $f \in L_{1,0}(A, \Sigma')$,*

$$\| Tf \| \geq \delta \| f \|_1.$$

In the particular case that E is a quotient L_1/X of L_1 and T is the quotient map:

Corollary 3.4.f. *They are equivalent:*

    i) the quotient map $L_1 \to L_1/X$ does not fix copies of L_1;
    ii) for all $\delta > 0$, $\Sigma' \subset \Sigma$ atomless and $A \in \Sigma'$ with $\lambda(A) \geq 0$ there exist $f \in L_{1,0}(A, \Sigma')$ and $g \in X$ such that
$$\| f \|_1 \geq \frac{1}{4} \quad and \quad \| f - g \|_1 \leq \delta.$$

The set up is completed when Talagrand [342, Prop. 4.4] shows:

Proposition 3.4.g. *Let $q: X \to Z$ be a quotient map. Assume that Z satisfies the following abstract condition: there is a constant $K > 0$ and there is a method Λ to associate to each sequence (y_n) of elements of some set $q^{-1}(z)) \cap Ball(X)$ with*

$\| z \| \leq 1$, *an element of X in such a way that:*

1) *if two sequences are eventually the same, their image by Λ is the same.*
2) $\| \Lambda((y_n)_n) \| \leq K.$
3) $q(\Lambda((y_n)_n)) = z.$
4) $\Lambda(\frac{1}{2}(x_n + y_n)_n) = \frac{1}{2}[\Lambda((x_n)_n) + \Lambda((y_n)_n)].$

If Z contains a copy of L_1 then q fixes a copy of L_1 .

Second portion: The construction of X_1 y X_2. The hard-analysis part of the proof is the construction of a sequence (C_n) of families of functions $C_n = (f_{n, k})_k$ so that

for every $\Sigma' \subset \Sigma$ atomless and for every $A \in \Sigma'$ with $\lambda A \geq 2^{-n}$ there exist $f \in L_{1, 0}(A, \Sigma')$ and $f_{n, k} \in C_n$ such that

$$\| f \|_1 \geq \tfrac{1}{4} \quad y \quad \| f - f_{n, k} \|_1 \leq 2^{-n}.$$

If H_n denotes the closed linear span of $(C_{n, k})_k$, set

$$X_1 = \overline{\text{span}} \ \bigcup_n H_{2n} \ \text{and} \ X_2 = \overline{\text{span}} \ \bigcup_n H_{2n+1}.$$

Proposition 4.1 of [342] shows that the spaces H_n are isomorphic to l_1 while Proposition 4.2 of [342] shows that the spaces X_1 and X_2 are isomorphic to l_1-sums of the corresponding H_n. This means that neither X_1 nor X_2 contain copy of L_1. That the product $L_1/X_1 \times L_1/X_2$ contains L_1 follows from the fact [342, p.21] that the natural map

$$L_1 \to L_1/X_1 \times L_1/X_2$$

given by $f \to (f + X_1, f + X_2)$ is an into isomorphism.

Third portion: The end. *The quotient spaces L_1/X_1 and L_1/X_2 do not contain L_1.*

In what follows we write X to denote X_1 or X_2. Consider the following diagram

$$0 \to X \to L_1 \to L_1/X \to 0$$
$$j\downarrow \qquad \downarrow \qquad \|$$
$$0 \to l_1 \to O \to \Sigma/l_1 \to 0$$
$$p$$

where $j: X \to l_1$ is a suitable defined map and O is the push-out of j and the inclusion $i: X \to L_1$. As we know, L_1/X is isomorphic to O/l_1. Talagrand shows that O/l_1 does not contain L_1 by proving:

i) O does not contain L_1; and
ii) the quotient map $p: O \to O/l_1$ verifies conditions 1-4 of 3.4.g. □

3.5   3*SP* problems for C(K) spaces

Three-space problems for $C(K)$ spaces seems to be a topic widely open. The following result is classical.

Theorem 3.5.a. *Not containing an isomorphic copy* $C[0, 1]$ *is a 3SP property.*

Proof. The proof follows and analogous schema as for l_∞. Let $0 \rightarrow Y \rightarrow X \rightarrow Z \rightarrow 0$ be a short exact sequence and let $L: C[0, 1] \rightarrow X$ be an isomorphic embedding. In this case no Diestel-Faires result is available since $C[0, 1]$ admits spaces such as c_0 as quotients. The proof therefore relies on a property of the operator qJ different from weak compactness. Consider first the space $C[0, 1]$ written as $C(\Delta)$, where Δ denotes the Cantor set. Let $T: C(\Delta) \rightarrow Z$ be a continuous operator. Here is the key property:

(*) For every $\varepsilon > 0$ and for every clopen nonempty subset $\Lambda \subset \Delta$ there is some $f \in C(\Lambda)$ with $\|f\| = 1$ and $\|Tf\| \leq \varepsilon$.

Property (*) is strictly weaker than the property of not being invertible on any infinite dimensional subspace of $C(\Delta)$, but it is however strong enough to guarantee the possibility of producing, for every $\varepsilon > 0$, an (isometrically) equivalent copy (e_n) of the Haar basis of $C(\Delta)$ so that $\Sigma_n \|Te_n\| < \varepsilon$ [246, Prop. 2.1].

Now, either qJ satisfies (*) or not. If it does not, qJ is clearly an isomorphism when restricted to $C(\Lambda)$ for some clopen set Λ. If qJ satisfies (*), it is possible to select (e_n) so that (Je_n) is a basic sequence in X equivalent to the Haar basis of $C(\Delta)$, with basis constant K, and such that $\Sigma \|qJe_n\| < (2K)^{-1}$. Therefore, it is possible to find elements (y_n) such that $\Sigma \|y_n - Je_n\| \leq (2K)^{-1}$. The sequence (y_n) is a basic sequence in Y equivalent to (e_n) and thus spanning $C(\Delta)$. $\square$

This result can be found in [246], where the 3*SP* problem for "*not containing* L_1" was raised.

Problems. The 3*SP* problem for $C(K)$ spaces can be formulated as: *Is every extension of* $C(K_1)$ *and* $C(K_2)$ *a space* $C(K)$ *for some compact space* K? *If so, when is* $K = K_1 \cup K_2$? *When does a twisted sum of* $C(K)$ *spaces split?*

A basic technique to construct compact spaces is described now.

Stone compacts. Let Ξ be an algebra of subsets on $\mathbb{N}$ containing the class $P_f(\mathbb{N})$ of all finite subsets of $\mathbb{N}$. Let K_Ξ be the subspace of $\{0, 1\}^\Xi$ (product topology) formed by all Boolean homomorphism; i.e., applications $h \in \{0, 1\}^\Xi$ such that

$$h(A) \qquad = 1-h(\mathbb{N}-A)$$
$$h(A\cup B) = max\{h(A), h(B)\}$$
$$h(A\cap B) = min\{h(A), h(B)\}.$$

The space K_Ξ is closed since all conditions are, and thus K_Ξ is a compact space whose topology is generated by the clopen sets $A \in \Xi$. By Stone's representation theorem, the space $C(K_\Xi)$ is generated by the characteristic functions of the sets in Ξ.

Identifying each $n \in \mathbb{N}$ with the Boolean homomorphism $\delta_n \in K_\Xi$ given by $\delta_n(A)=1$ if $n \in A$ and $\delta_n(A)=0$ if $n\notin A$, one obtains a copy of $\mathbb{N}$ in K_Ξ. This copy is dense: given $A_1,\dots,\ A_n \in \Xi$ defining an open set $V=\{\delta \in K_\Xi : \delta(A_i)=1,\ 1\leq i\leq n\}$ (if the condition is $\delta(A)=0$ then it can be replaced by $\delta(\mathbb{N}\backslash A)=1$) one must have $A_1\cap\dots\cap A_n \neq \varnothing$; if k is a point in the intersection then $\delta_k \in V$. Therefore K_Ξ can be considered as a compactification of $\mathbb{N}$.

Moreover, the embedding $c_0 \to C(K_\Xi)$ given by $e_n \to 1_{\delta_n}$ is an isometry into. We shall always refer to this copy of c_0. It is not difficult to see that there is an isometry $C(K_\Xi)/c_0 = C(K_\Xi\backslash\mathbb{N})$. Thus, there is an exact sequence

$$0 \to c_0 \to C(K_\Xi) \to C(K_\Xi\backslash\mathbb{N}) \to 0.$$

The 3SP problem for C(K) spaces.
The $3SP$ problem for $C(K)$ spaces has a negative answer as it was observed by F. Cabello [45].

Theorem 3.5.b. *To be isomorphic to a $C(K)$-space is not a 3SP property.*

Proof. Let $\phi\colon \beta\mathbb{N}-\mathbb{N} \to \beta\mathbb{N}-\mathbb{N}$ be an identification of the natural copy of $\beta\mathbb{N}-\mathbb{N}$ inside $\beta\mathbb{N}$ with $\beta\mathbb{N}-\mathbb{N}$. Let K the disjoint union of $\beta\mathbb{N}$ and $\beta\mathbb{N}-\mathbb{N}$. Given $0<\tau<1$, let

$$X_\tau = \{f \in C(K) : f(U) = \tau f(\phi(U)) , \ U \in \beta\mathbb{N}-\mathbb{N} \}.$$

This construction is due to Benyamini [22], who shows that $X_{1/\tau}$ is not θ-isomorphic to a complemented subspace of a $C(K)$ space in the following sense: if $T\colon X_{1/\tau} \to C(K)$ is an into isomorphism and P is a projection of $C(K)$ onto $TX_{1/\tau}$ then $\|T\|\ \|T^{-1}\|\ \|P\| \geq 1/\tau$. Obviously, X_τ is $1/\tau$-isomorphic to l_∞.

Let

$$X = c_0 (X_1, X_{1/2}, \dots , X_{1/n},\dots).$$

After the result of Benyamini it is clear that X is not isomorphic to a complemented subspace of a $C(K)$-space. Observe that c_0 is isometric to the natural copy of c_0 inside X_τ, and in such a way that X_τ/c_0 is isometric to $C(\beta\mathbb{N}-\mathbb{N})$. Thus,

$c_0 = c_0(c_0)$ is isometric to a subspace of X in such a way that X/c_0 is isometric to $c_0(l_\infty/c_0)$, which is isomorphic to $C(\mathbb{N}^* \times (\beta\mathbb{N} - \mathbb{N}))$. Therefore, there exists a short exact sequence

$$0 \to c_0 \to X \to C(\mathbb{N}^* \times (\beta\mathbb{N} - \mathbb{N})) \to 0$$

where X is not isomorphic to a $C(K)$ space.   □

The algebraic part of the construction is simple; if V_θ is the pull back of the quotient map $q: l_\infty \to l_\infty/c_0$ and a dilation $\theta id: l_\infty/c_0 \to l_\infty/c_0$ then one obtains the diagram

$$0 \to c_0 \to l_\infty \to l_\infty/c_0 \to 0$$

$$\| \qquad \uparrow \qquad \uparrow \theta \, id$$

$$0 \to c_0 \to V_\theta \to l_\infty/c_0 \to 0$$

which shows that V_θ is θ^{-1}-isomorphic to l_∞. The space $X = c_0(V_1, \ldots, V_{1/n}, \ldots)$ provides the desired exact sequence.

Moreover, since V_θ is a pull-back, the sequence

$$0 \to V_\theta \to c_0 \times l_\infty/c_0 \to l_\infty/c_0 \to 0$$

is also exact. Therefore there is an exact sequence

$$0 \to X \to C(K) \to C(D) \to 0$$

where X is not a $C(K)$-space.

Several partial results that can be mentioned: *i*) *If K is a metrizable compact space, any short sequence* $0 \to c_0 \to X \to C(K) \to 0$ *splits* (Sobczyk, see [354]). *ii*) *The sequence* $0 \to C[0, \omega_1] \to C([0, \omega_1] \cup [0, \omega_1]) \to C[0, \omega_1] \to 0$ *splits* (although $C[0, \omega_1]$ and $C[0, \omega_1] \times C[0, \omega_1]$ are not isomorphic) (Semadeni, see also 3.6.b in appendix 3.6). *iii*). If $K=[0, 1] \times \{0, 1\}$ with the lexicographical order topology then $C(K)=D[0, 1]$. *Any exact sequence* $0 \to c_0 \to C(K) \to Z \to 0$ *with i isometry into, splits and is equivalent to* $0 \to c_0 \to C(K) \to C(K) \to 0$ ([278]). *iv*) Let D be a metrizable subspace of a compact space K and let $J=\{f \in C(K) : f(D)=0\}$. *Every short exact sequence* $0 \to J \to C(K) \to C(D) \to 0$ *splits* (Pelczynski, see [162]). Even when D is not metrizable the dual sequence splits (and with the l_1-norm; i.e., J is an M-ideal of $C(K)$; see [162] for *all* information about M-ideals).

Appendix 3.6

Banach spaces not isomorphic to their square

Banach posed in [15] the following questions:

i) Is every infinite dimensional Banach space X isomorphic to its product $X \times \mathbb{R}$ by the real line?

ii) Is every infinite dimensional Banach space X isomorphic to its square $X \times X$?

The first question asks if every closed subspace of codimension one of X is isomorphic to X and so it has been known as *the hyperplane problem*. The solution to this problem did not come until recently when Gowers [154] answered it in the negative. Let us first make some comments on the second question.

Two negative answers appeared to the second question almost simultaneously in [24] and [326]. Those examples are interesting for us because given a space X not isomorphic to its square, the class *Banach spaces isomorphic to X* fails the 3SP property in a rather strong way. Trivial examples of Banach spaces not isomorphic to their square are finite dimensional spaces (in fact, some of the infinite dimensional examples are based in this one). It is an obvious observation that if X^* (or X^{**}/X) is not isomorphic to its square, neither is X.

Proposition 3.6.a. [24] *If X is quasi-reflexive, non-reflexive and Y is weakly sequentially complete then X and $X \times Y$ are not isomorphic to their square.*

Proof. The case of X follows immediately from the fact that X^{**}/X is finite dimensional, hence, X is not isomorphic to its square. For the other case, observe that if $B_1(X)$ denote as in 2.3.b the subspace of those elements of X^{**} that are weak*-limits of sequences of elements of X then

$$dim\ B_1(X \times Y)/X \times Y = dim\ X^{**}/X. \qquad \square$$

With this, one obtains as in [24] that the following spaces are not isomorphic to their squares: James' space J, since it is quasi-reflexive; $J \times C[0, 1]$ since J^* is quasi-reflexive and $C[0, 1]^*$ is weakly sequentially complete. Spaces X such that X^{**}/X is isomorphic J cannot be isomorphic to their square. Another example more elaborate appears in [326]:

Proposition 3.6.b. *The space $C[0, \omega_1]$ is not isomorphic to its square.*

Proof. Let

$$S_1(X) = \{ F \in X^{**} : F \text{ is weak* sequentially continuous}\};$$

equivalently, the kernel of F is weak* sequentially closed. It will be enough to show that $S_1(C[0, \omega_1])/C[0, \omega_1]$ is one-dimensional. Note that $C[0, \omega_1]^* = l_1[0, \omega_1]$ and thus $C[0, \omega_1]^{**} = l_\infty[0, \omega_1]$.

Claim: The element $1_{\omega_1} \in C[0, \omega_1]^{**}$ defined by $1_{\omega_1}(f) = f(\omega_1)$ belongs to $S_1(C[0, \omega_1])$.

Proof of the Claim. If $f = (f(\alpha))_\alpha \in l_1[0, \omega_1]$ then $supp(f) = \{\alpha : f(\alpha) \neq 0\}$ is countable; hence, $supp(f) \subset [0, \alpha]$ for some $\alpha < \omega_1$. The same happens for elements $f \in Ker(1_{\omega_1})$; thus, the union of the supports of a sequence of elements in $Ker(1_{\omega_1})$ is countable and $Ker(1_{\omega_1})$ is weak* closed. $\square$

Second Claim: Every $F \in S_1(C[0, \omega_1])$ is continuous on $[0, \omega_1)$.

Proof of the Second Claim. If $F = (F(\alpha))_\alpha \in l_\infty[0, \omega_1]$ is not continuous on $[0, \omega_1)$, there would be an increasing sequence $\beta_1 < \beta_2 < \ldots < \beta_n < \ldots$ convergent to some $\beta < \omega_1$ and such that

$$lim_{n \to \infty} F(\beta) = L \neq F(\beta).$$

There is no loss of generality assuming $L=0$ and $F(\beta)=1$; if otherwise, replace F by $F+\beta 1$. Take the elements $f_m \in l_1[0, \omega_1]$ given by

$$f_m(\beta_n) = \delta_{mn}, \text{ for } n=1, 2, \ldots \text{ and } f_m(\alpha) = 0, \text{ otherwise.}$$

The sequence (f_m) is weak* convergent to the characteristic function of β; i.e., the functional f such that $f(\beta)=1$ and $f(\alpha)=0$, otherwise. Moreover,

$$lim_{n \to \infty} F(f_n) = L \neq F(f) = 0.$$

Hence, $F \notin S_1(C[0, \omega_1])$. $\square$

End of the proof of the proposition. For $F \in S_1(C[0, \omega_1])$ the limit

$$lim_{\alpha \to \omega_1} F(\alpha)$$

exists, since otherwise there would be an increasing and convergent sequence (β_n)

in $[0, \omega_1)$ such that $(F(\beta_n))_{n \in \mathbb{N}}$ has no limit, in contradiction with the second claim.

Define $G(\alpha) = F(\alpha)$ for $\alpha < \omega_1$ and $G(\omega_1) = lim_{\alpha \to \omega_1} F(\alpha)$. One has $G \in C[0, \omega_1]$ and

$$F = G + (F(\omega_1) - G(\omega_1)) 1_{\omega_1}.$$

The proposition is proved.   □

A (more complicated) way to obtain spaces not isomorphic to its square is obtaining *real* Banach spaces admitting no complex structure. This has been done by Szarek [335], showing that there exists an infinite dimensional superreflexive real Banach space that does not admit complex structure. In particular, it is not isomorphic to the square of any Banach space. Gowers and Maurey [157] show that there exists a Banach space X isomorphic to $X \times X \times X$, but not isomorphic to $X \times X$!

A related problem is that of the existence of real-isomorphic Banach space that are not complex isomorphic. Kalton in an unpublished manuscript solved the corresponding isometric problem; i.e., he showed the existence of real isometric non complex-isometric spaces. Bourgain [35] solved the isomorphic problem constructing two real-isometric spaces not complex isomorphic. A method using twisted sums to obtain *real* isomorphic spaces that are not complex isomorphic was provided by Kalton [209]:

Sketch of the construction. First of all, observe that twisted sums can be translated to complex spaces without further difficulties. If $|\alpha| < 1$ is a real number and $f_\alpha(t) = t^{1+i\alpha}$ and $Z_2(\alpha)$ denotes the twisted sum $l_2 \oplus_F l_2$ constructed with the quasi-linear map induced by f_α then a combination of the following simple result

Proposition 3.6.c. *The complex conjugate of $Z_2(\alpha)$ is isomorphic to $Z_2(-\alpha)$.*

and the following non-trivial one

Proposition 3.6.d. *$Z_2(\alpha)$ and $Z_2(\beta)$ are isomorphic if and only if $\alpha = \beta$.*

yields the existence of a real Banach space not isomorphic to its complex conjugate.

About $3SP$ questions, Kaibkhanov proves in [201] that there exist Banach spaces Y and Z not isomorphic to their square such that $Y \times Z$ is isomorphic to its square (which also follows from Gowers-maurey space X just mentioned: X is not isomorphic to $X \times X$ while $X \times X \times X \times X$ is isomorphic to $X \times X$ since X is isomorphic to $X \times X \times X$. Therefore,

Theorem 3.6.e. *The class of Banach spaces not isomorphic to their square has not*

the 3SP property

Returning to the hyperplane problem, it is clear that none of the ideas that worked to distinguish X from $X \times X$ is fine enough to distinguish between X and $X \times \mathbb{R}$. The example of Gowers [154] is an adaptation of the space of Gowers-Maurey [156]: while the latter does not admit unconditional basic sequences, the former has an unconditional basis. With this setting, the basic tool is the following result of Casazza [154, Lemma 9].

Lemma 3.6.f. *Let X be a Banach space with a basis with the property that whenever (y_n) and (z_n) are two sequences such that $y_n < z_n < y_{n+1}$ (with respect to the basis), they are not equivalent. Then X is not isomorphic to any proper subspace.*

Thus, the (nontrivial) task of Gowers is to verify that a certain Schlumprecht-like space X with basis can be constructed satisfying the assumptions of the lemma.

Appendix 3.7

Dual Banach spaces

Is the property of being isomorphic to a dual space a 3SP property? A variation of this problem seems to have been suggested by Vogt [355]. Precisely, Vogt's question seems to be: *if one has a short exact sequence*

$$0 \to Y^* \to X \to Z^* \to 0,$$

must this be a dual sequence?

In the category of Fréchet spaces the answer is no, as the following example of Díaz, Dierolf, Domanski and Fernández shows [84]: Let E be a Fréchet Montel space having a quotient isomorphic to l_1 (see [225]) and let M be the kernel of any quotient map $l_1 \to L_1(0, 1)$; M is not complemented in its bidual (see [243]). If $q: E \to l_1$ is the quotient map and $i: M \to l_1$ the inclusion, the pull back Ξ of $\{q, i\}$ determines an exact sequence $0 \to \Xi \to M \oplus E \to l_1 \to 0$. The pull-back Ξ is a dual space since it is isomorphic to $q^{-1}(M)$, a subspace of a Montel space, again Montel and thus reflexive. The space $M \times E$ is not a dual because it is not complemented in its bidual. ☐

In the category of Banach spaces the answer is again no. The construction of a counterexample in full details can be seen in [48]; it is based on a deep development of the theory of 0-linear maps. We present however an sketch of the construction. First, let us show a partial positive result (see [84]; and also [48]).

Proposition 3.7.a. *If* $0 \to Y \to X \to Z \to 0$ *is a short exact sequence where Z is reflexive and Y is a dual space then X is a dual space and the sequence is a dual sequence.*

Proof. Since $0 \to Y \to X \to Z \to 0$ is exact, the dual sequence is also exact. Let $_*Y$ denote the predual of Y. If $W = (i^*)^{-1}(_*Y)$ and $j = (i^*_{|W})^*$, then the diagram

$$
\begin{array}{ccccccccc}
0 & \to & Y & \to & X & & \to & Z & \to & 0 \\
& & \| & & \downarrow & & & \| id & & \\
0 & \to & (_*Y)^* & \to & W^* & & \to & Z & \to & 0 \\
& & & j & & q^{**} & & & &
\end{array}
$$

shows that those two sequences are equivalent.   □

Theorem 3.7.b. *To be isomorphic to a dual space is not a 3SP property.*

Proof. In [215], it is shown the existence of a nontrivial exact sequence

$$0 \to l_2 \to X \to c_0 \to 0.$$

The bidual sequence

$$0 \to l_2 \to X^{**} \to l_\infty \to 0$$

does not split and thus the dual (of the starting) sequence

$$0 \to l_1 \to X^* \to l_2 \to 0$$

does not split. Let W be a Banach space such that $W^{**} = W \oplus l_1$ (see e.g., [176, 244]). The sequence

$$0 \to W \oplus l_1 \to W \oplus X^* \to l_2 \to 0$$

does not split. By 3.7.a, this sequence is dual of some sequence

$$0 \to l_2 \to {}_*(W \oplus X^*) \to W^* \to 0.$$

Moreover, the theory developed in [48] shows that this sequence cannot be the dual sequence of a sequence

$$0 \to W \to {}_{**}(W \oplus X^*) \to l_2 \to 0;$$

and thus, by results of Brown and Ito [42] (W^* has W as its unique isomorphic predual) it is not dual of any sequence. This answers Vogt's question in the negative. The middle space in the sequence

$$0 \to l_2 \to {}_*(W \oplus X^*) \to W^* \to 0$$

is not a dual space: otherwise, l_2 would be weak*-closed and the bipolar theorem would imply that it is a dual sequence.   □

Theorem 3.7.c. *To be complemented in its bidual is not a 3SP property*

Proof. See [48]. Let JT denote the James-Tree space and B its predual (see the example at 4.14.e for all pertinent information); the space B is uncomplemented in its bidual JT^* and one has the exact sequence

$$0 \to B \to JT^* \to l_2(I) \to 0.$$

Let $q: l_1(J) \to JT^*$ be a quotient map and $i: B \to JT^*$ the canonical inclusion. Let Ξ be the pull-back of the operators $\{q, i\}$. One has the commutative diagram

$$
\begin{array}{ccc}
0 & & 0 \\
\uparrow & & \uparrow \\
l_2(I) & = & l_2(I) \to 0 \\
\uparrow & & \uparrow \\
0 \to Y \to l_1(J) \overset{q}{\to} JT^* \to 0 \\
\parallel \qquad \uparrow \qquad \uparrow i \\
0 \to Y \to \Xi \to B \to 0 \\
\uparrow \qquad \uparrow \qquad \uparrow \\
0 \qquad 0 \qquad 0
\end{array}
$$

The pull-back space Ξ is complemented in its bidual since it is the kernel of a quotient map $q\colon l_1(J) \to l_2(I)$ (see [19]). Moreover, since q is surjective one has the exact sequence $0 \to \Xi \to l_1(J) \oplus B \to JT^* \to 0$. □

The proof of the following partial affirmative answer is just painting a full diagram.

Proposition 3.7.d. *Let* $0 \to Y \to X \to Z \to 0$ *be a short exact sequence. If Y is complemented in Y^{**} and Z is reflexive then X is complemented in X^{**}.*

Isometric preduals

A Banach space U is said to be a unique predual if $U^* = V^*$ implies $U = V$ isometrically. Godefroy asks in [133, (6) p.186] if a Banach space without copies of c_0 must be a unique predual. Also, Godefroy [133, (6) p.187] asks:

Problem. *If Y and X/Y are unique preduals, must X be a unique predual?*

A variation about this problem is: a Banach space is said to have *property* (X) (Saab and Saab [317, p.372]) if

$$Y = \{z \in Y^{**} : \text{ for all weakly-1-summable sequences } (f_n) \subset Y^*$$

$$z(weak^* \textstyle\sum_n f_n) = \textstyle\sum_n z(f_n)\}.$$

Problem. *If Y and X/Y are unique preduals and have property (X), must X be a unique predual?*

Saab and Saab mention that property (X) implies property (V^*) and cite, as examples of spaces with property (X) the following: $\mathcal{L}_1$-spaces; L_1/H_1; spaces with l.u.s.t. and not containing l_∞^n uniformly.

Chapter 4

Topological properties of Banach spaces

By *topological properties* we understand properties defined in topological terms (usually referred to the weak topology) that do not, in principle, depend of the linear structure of the space. They can be divided in two types: those derived from reflexivity when thought as the property that bounded sequences admit weakly convergent subsequences, and those derived from reflexivity when thought as the property that the topological space $(X, weak)$ is a countable union of compact sets.

Extraction of subsequences. Several interesting properties of Banach spaces can adopt an extraction-of-subsequences aspect: every sequence of a certain type **A** admits a subsequence of another type **B**. When property P is "bounded sequences admit subsequences of type B" then the property "weakly null sequences admit subsequences of type B" is usually called the *weak P* property.

Chronologically, the first results of this kind were established by Riesz: finite-dimensional spaces are those where bounded sequences admit convergent subsequences; and by Eberlein and Smullyan: reflexive spaces are those where bounded sequences admit weakly convergent subsequences. Related results are the proofs that every closed linear subspace of $C[0,1]$ is weakly closed (Zalcwasser) and that every closed convex subset of a Banach space is weakly closed (Mazur). This last result implies that every weakly convergent sequence admits a sequence of convex combinations norm convergent to the same limit.

Banach and Saks proved [16] that in the case of L_p-spaces the convex combinations can be chosen to be the arithmetic means. Accordingly, a Banach-Saks sequence is a sequence having norm convergent arithmetic means and the Banach-Saks property is the property that every bounded sequence admits Banach-Saks subsequences. Schreier [325] gave the first example (the so-called Schreier space) of a Banach space where some weakly null sequence does not contain Banach-Saks subsequences. In fact, this sequence could well be understood as a sequence of continuous functions on the compact space ω^ω. The next breakthrough was Kakutani's proof [202] that a superreflexive space has the Banach-Saks property. This led Sakai [319] to ask whether every space with the Banach-Saks property had to be reflexive, a question that found affirmative answer with the proof of Nishiura and Watermann [267].

The p-Banach-Saks property was introduced by Johnson [190, 191] when studying the factorization of operators through l_p. In fact, the property Johnson introduced was the weak p-Banach-Saks property (every weakly null sequence admits a subsequence having bounded p-means). The W_p properties were introduced in [67] trying to fill the gap between super-reflexivity and reflexivity by asking that every bounded sequence admits a subsequence weakly-p-convergent. The weak versions (weakly null sequences admit weakly p-summable subsequences) have also interest: for instance, *weak W_1* coincides with the *hereditary Dunford-Pettis property*. These weak versions were also introduced by Knaust and Odell [223] studying the presence of sequences equivalent to the canonical basis of l_p. Kakutani's theorem can be improved to *superreflexivity* → W_p (for some p). To complete the picture, W_p obviously implies *p^*-Banach-Saks*, and it can be proved [67] that p^*-Banach-Saks implies W_r for all $r > p$. The same happens with their weak versions. Rakov [303] gave an example showing that *weak p^*-Banach-Saks* does not imply *weak W_p*.

Weaker variations of reflexivity are the quasi-reflexivity after James [174], the co-reflexivity of Clark [71] and somewhat reflexivity.

One has the following schema:

$$super\text{-}reflexive \Rightarrow W_p \Rightarrow p^*\text{-}BS \Rightarrow BS \Rightarrow reflexive$$
$$\Downarrow \qquad\qquad\qquad \Downarrow$$
$$W_r \quad \forall\, r > p \qquad quasi\text{-}reflexive$$
$$\Downarrow$$
$$co\text{-}reflexive$$
$$somewhat\ refleixve$$

All implications are proper (see [67] and 4.4) except p^*-Banach-Saks $\Rightarrow W_p$ which is open.

The property of having weakly* sequentially compact dual ball plays an important role in Banach space theory, though there are many obscure points about it. Its $3SP$ properties shall be considered here and in Chapter 6.

The topological space (X, weak). Recall that if A denotes a certain class of subsets of a space then A_σ is the class of those sets that are countable unions of elements of A, while A_δ are countable intersections of elements in A; analogously, $A_{\sigma\delta}$ means countable unions of countable intersections of elements in A; the letter K is reserved for the class of compact sets. From this point of view, Baire's lemma yields *reflexivity* $=$ (X, weak) is a K_σ. A weakening of reflexivity is the notion of *weakly compactly generated (WCG)* space: (X, weak) admits a *dense K_σ*. Still, they can be considered weaker properties (precise definitions shall be given below) such as: X is a $K_{\sigma\delta}$ of $(X^{**},\ weak^*)$; (X, weak) is the continuous image of a $K_{\sigma\delta}$ in some compact space (X is *weakly K analytic*); points of X are covered, on an individual basis, by countable intersections of weakly* compact sets of X^{**} whose intersection is contained in X (*weakly countably determined* or *WCD*); there is a transfinite chain of projections whose images eventually cover X (X admits a *projectional resolution of the identity* or *PRI*). These and many other related properties shall be considered in this chapter. It is a good warning to remark that papers [106, 107, 108, 187, 188, 296, 315, 338, 352, 357] contain many more properties of topological nature. We just highlight here those for which there are $3SP$ results available. The following schema pictures the relationships among those notions.

$$
\begin{array}{l}
\textit{reflexive} \;\Rightarrow\; \textit{quasi-reflexive} \\[4pt]
\quad\Downarrow \qquad\qquad \Downarrow \\[4pt]
\textit{Cech}^{\,dual} \;\Leftrightarrow\; \textit{Sep}^{\,co} \;\Leftrightarrow\; \textit{PV}^{\,dual\,dual} \;\Rightarrow\; \textit{somewhat reflexive} \\[4pt]
\quad\Downarrow \;\neq \\[2pt]
\textit{Polish} \;\Rightarrow\; \textit{Cech} \;\Rightarrow\; \textit{PCP} \\[4pt]
\qquad\qquad \Downarrow \\[2pt]
\textit{S.D.} \;\Rightarrow\; \textit{PV}^{\,dual} \;\Rightarrow\; \textit{WCG}^{\,dual} \;\Rightarrow\; \textit{Asplund} \;\Rightarrow\; \textit{weak Asplund} \\[2pt]
\qquad\quad \neq \\[2pt]
\quad\Downarrow \\[2pt]
\textit{PV} \;\Rightarrow\; \textit{WCG} \;\Rightarrow\; \textit{weak}^{*}K_{\sigma\delta} \;\Rightarrow\; \textit{wK} \quad \textit{analytic} \quad \;\Rightarrow\; \textit{WCD} \;\Rightarrow\; \textit{PRI} \\[4pt]
\qquad\qquad\qquad\qquad\qquad\qquad \Downarrow \\[2pt]
\qquad\qquad\qquad\qquad\qquad \sigma\textit{-fragmentable}
\end{array}
$$

A different door to the understanding of topological properties in terms of linear structure of the space was opened by Valdivia's paper [349] where it is shown that *Separable ^{co}* implies *Reflexive $\oplus$ Separable*. Spaces admitting this decomposition were also pointed out by Plichko [295]; so, we define the *Plichko-Valdivia* property (*PV*) as *Reflexive $\oplus$ Separable*. Following this approach, spaces in *Reflexive $\oplus$ Separable ^{dual}* are called *separately distinguished* (*SD*). A thorough study of *P-by-Q* and *P $\oplus$ Q* properties, called *twisted properties*, can be found in [63].

4.1   Reflexivity

A Banach space X is said to be *reflexive* if the canonical injection $X \to X^{**}$ is surjective; equivalently, the closed unit ball is a weakly compact set.

Theorem 4.1.a. *Reflexivity is a 3SP.*

This result was first proved by Krein and Smullyan [226]. Their proof was made for what they called "regular" spaces i.e., spaces such that for each $x^{**} \in X^{**}$ some $x \in X$ exists such that $x^{**}(f)=f(x)$ for all $f \in X^{*}$.

1st *Proof.* Let $0 \to Y \to X \to Z \to 0$ be a short exact sequence with both Y and Z reflexive. Let $L: Z \to X$ be a linear (possibly discontinuous) selection for q, and let $x^{**} \in X^{**}$. Since $q^{**}x^{**} \in Z^{**} = Z$, The difference $x^{**}-L(q^{**}x^{**}) \in Y^{**}$ because it acts linear and continuously on Y^{*}; it can thus be represented by some element $y \in Y$. Clearly, $x^{**} = L(q^{**}x^{**})+y \in X$.   □

2nd *Proof.* As it was already mentioned in 2.3.b, the 3SP property of reflexivity is a consequence of the exactness of the functor $X \to X^{**}$.   □

This way of proof, and with it the appearance of short exact sequences in Banach space theory, is due to Yang [359]. Later, Webb [70] uses the same ideas to prove that quasi-reflexivity and co-reflexivity are 3SP properties. Yang ([360] and [361]) introduces new elements from homological algebra that applies to the study of Fredholm operators. It is perhaps worth to remark that paper [1] is an interesting attempt to introduce new elements of homological algebra into Banach space (algebra) theory. Precisely, after rescuing the push-out construction (lemma 6.6) they define c_0-*ext* spaces as spaces admitting a type of presentation of length n. They prove the following three-space-type result:

> Let $0 \to Y \to X \to Z \to 0$ be a short exact sequence where Z is realcompact and admits a C^{∞}-partition of unity and Y is in c_0-ext. Then all algebra homomorphisms of $C^{\infty}(X)$ are point evaluations.

4.2   Quasi-reflexivity

James [174] was able to construct a Banach space J such that, being isometric to J^{**}, its canonical image in J^{**} is of codimension 1. A Banach space X is said to be *quasi-reflexive* if $qr(X)=dim(X^{**}/X) < \infty$. Civin and Yood [70] define a Banach space X to be quasi-reflexive of order n if $qr(X)=n$. They prove that

$$qr(X) = qr(Y) + qr(X/Y).$$

This result also gives the $3SP$ property for quasi-reflexivity and reflexivity.

4.3   Co-reflexivity

Clark defines in [71] a Banach space X to be *coreflexive* if X^{**}/X is reflexive; i.e., if $X \in Reflexive^{co}$. X is said to be *complemented coreflexive* if it is complemented in X^{**} by a reflexive subspace. These two notions are different: the predual B of James-Tree space JT (see 4.14.e) is uncomplemented in JT^* while $JT^*/B = l_2(I)$. Since the functor ^{co} is exact, one has:

Theorem 4.3.a. *Co-reflexivity is a $3SP$ property.*

Question. *Is complemented coreflexivity a $3SP$ property?*

4.4   Somewhat reflexivity

The Banach space X is said to be *somewhat reflexive* if every closed infinite dimensional subspace contains an infinite dimensional reflexive subspace. Applying the incomparability method one has:

Theorem 4.4.a. *Somewhat reflexivity is a $3SP$ property.*

A somewhat reflexive space need not be quasi-reflexive: Herman and Whitley prove in [168] that James quasi-reflexive space J is l_2-saturated, from where they obtain that $l_2(J)$ is l_2-saturated although not quasi-reflexive since $l_2(J)^{**}/l_2(J) = l_2$.

A related result obtained by Clark is: If $X \in Separable^{co}$, i.e., if X^{**}/X is separable then X and X^* are somewhat reflexive. Moreover, by a result of Valdivia [349] (see 4.10.i) spaces in $Separable^{co}$ can be decomposed into a direct sum of a reflexive and a separable space (see also 4.14.g).

4.5   Super-reflexivity

A Banach space X is said to be *super-reflexive* if every Banach space finitely representable in X is reflexive. Recall that a Banach space E is said to be finitely representable in X if given $\varepsilon > 0$ and a finite dimensional subspace F of E there is a finite dimensional subspace G of X which is $(1+\varepsilon)$-isomorphic to F. Typical examples of spaces finitely represented in X are X^{**} or ultrapowers X_U of X.

Since finite dimensional subspaces of E carry a natural partial order, taking a nontrivial ultrafilter supported by that order one can prove that E is finitely representable in X if and only if E is isometric to a subspace of some ultrapower of X. So, a space is superreflexive if and only if every ultrapower is reflexive. Therefore, since the functor ^{up} is exact, applying the method 2.2.h one obtains:

Theorem 4.5.a. *Super-reflexivity is a 3SP property.*

Super-reflexive spaces enjoy some additional properties regarding extraction of subsequences, as we shall explore now. For instance, by a result of Enflo [114], superreflexive spaces are those admitting an equivalent uniformly convex norm; and Kakutani proved that uniformly convex spaces have the so-called Banach-Saks property (see below).

4.6 W_p and p-Banach-Saks properties

A sequence (x_n) in a Banach space X is said to be *weakly-p-summable*, $p \geq 1$, if for every $x^* \in X^*$ the sequence $(x^*(x_n))_n \in l_p$; equivalently, there is a $C > 0$ such that

$$\left\| \sum_k \xi_k x_k \right\| \leq C \|(\xi_k)\|_{p^*}$$

for any $(\xi_n) \in l_{p^*}$. It is said to be *p-Banach-Saks* (see [190, 191]), $1 < p < +\infty$, if for some constant $C > 0$ and all $n \in \mathbb{N}$

$$\left\| \sum_{k=1}^{n} x_k \right\| \leq C n^{1/p}.$$

The sequence is said to be a *Banach-Saks sequence* if it has norm convergent arithmetic means. We shall say that the sequence (x_n) is *weakly-p-convergent* (resp. *p-Banach-Saks convergent*) to $x \in X$ if the sequence $(x_n - x)$ is weakly-p-summable (resp. p-Banach-Saks). Obviously weakly-p-summable sequences are p^*-Banach-Saks. The converse is, in general, false: the sequence $(n^{-1/2})$ is 2-Banach-Saks in $\mathbb{R}$, but it is not 2-summable.

A subset is said to be *weakly-p-compact*, $1 \leq p \leq \infty$, if every sequence admits a weakly-p-convergent subsequence. A Banach space X is said to belong to W_p if its closed unit ball is weakly-p-compact. It is said to have the *p-Banach-Saks property* [190, 191] if any bounded sequence admits a p-Banach-Saks convergent subsequence. It is said to have the *Banach-Saks property* [16] (case $p=1$) if bounded sequences admit Banach-Saks subsequences. When "bounded sequence" is replaced by "weakly null sequence" we speak of those properties in the weak sense; thus, the weak Banach-Saks property means that every weakly null sequence admits a Banach-Saks subsequence, etc.

Kakutani proved in [202] that uniformly convex spaces have the Banach-Saks property; it is possible to prove (see [67]) that super-reflexive spaces belong to some class W_p. Other examples are: For $1 < p < \infty$, $l_p \in W_r$ if and only if $r \geq p^*$. For $1 < p < \infty$, $L_p(\mu) \in W_r$ if and only if $r \geq \max(2, p^*)$. Tsirelson dual space is such that $T^* \in W_p$ for all $p > 1$. One has:

Theorem 4.6.a. *The weak Banach-Saks, weak-W_p, $1 \leq p < +\infty$, and weak p-Banach-Saks properties, $1 \leq p < +\infty$, are not 3SP properties.*

Proof. Consider S to be Schreier's space [325], the first example of a space without the weak-Banach-Saks property. This is the space obtained by completion of the space of finite sequences with respect to the following norm:

$$\|x\|_S = \sup_{[A \ admissible]} \sum_{j \in A} |x_j|,$$

where a finite subset of natural numbers $A = \{n_1 < \ldots < n_k\}$ is said to be *admissible* if $k \leq n_1$.

It can be readily verified that S is algebraically contained in c_0, that it has an unconditional basis formed by the canonical vectors (e_i), and that a sequence (x^n) of S is weakly null if and only if, for every j, the sequence $(x_j^n)_{n \in \mathbb{N}}$ is null. Therefore, the canonical basis (e_n) is a weakly null sequence. Moreover, the sequence (e_n) does not admit Banach-Saks subsequences since the estimate

$$\left\| \sum_{j=1}^{N} e_{i_j} \right\| \geq \frac{N}{2}$$

holds.

The common counterexample shall be provided by the diagonal short exact sequence induced by the pull-back of two arrows: the canonical inclusion $i: S \to c_0$ and a quotient map $q: l_1 \to c_0$; that is

$$0 \to Ker(q+i) \to l_1 \oplus S \to c_0 \to 0.$$

It is only needed to verify that $Ker \ (p+i)$ has *weak* W_1, since c_0 obviously has it. To this end, let (y^n, x^n) be a weakly null sequence in $Ker \ (p+i)$. Since $py^n = -x^n$ and (y^n) is norm null, one sees that also $\| x^n \|_\infty \to 0$. If $\|x^n\|_S \to 0$, then the proof ends. If not, by the Bessaga-Pelczynski selection principle, there is no loss of generalityy assuming that the sequence (x^n) is formed by normalized blocks of the canonical basis (e_i):

$$x^n = \sum_{i=N_n+1}^{i=N_{n+1}} \lambda_i e_i .$$

Since $\|x^n\|_S = 1$ and $\|x^n\|_\infty \to 0$, one sees that $N_{n+1} - N_n \to +\infty$. Choose a subsequence $(x^m) = x^{k(m)}$ satisfying:

$$N_{k(m+1)} - N_{k(m)} > N_{k(m-1)} ,$$

and

$$\max_j |x_j^m| \le \min_j |x_j^{m-1}| ,$$

where the *min* is taken over the non-vanishing coordinates. One has

$$\| \sum_{m=1}^N x^m \|_S \le 2,$$

because if $A = \{ n_1 < \ldots < n_k \}$ with $k \le n_1$ is admissible, and $N_{k(m-1)} < n_1 \le N_{k(m)}$ then

$$\sum_{j\in A} |\sum_{m=1}^N x_j^m| \le \text{ sum of } k \text{ consecutive terms beginning with } x_{n_1}^m$$

$$\le \|x^m + x^{m+1}\|$$

$$\le 2. \qquad\qquad \square$$

This counterexample shall appear again in Chapter 6 as a counterexample to the 3*SP* problem for the Dunford-Pettis and hereditary Dunford-Pettis property. This constructions was first used by Ostrovskii [272] to prove that the weak-Banach-Saks is not a 3*SP* property. On the other hand, answering a question of Partington [277], Godun and Rakov [137] proved

Theorem 4.6.b. *The Banach-Saks property is a 3SP property.*

Recall that a Banach space X has the *Alternate Banach-Saks property* if bounded sequences (x_n) admit subsequences (x_m) such that $\sum_m (-1)^m x_m$ converges. We shall present the proof of Ostrovskii [271], who derived 4.6.b from the equivalence *Banach-Saks = Alternate Banach-Saks + Reflexivity* (see [20]) and

Theorem 4.6.c. *The alternate Banach-Saks property is a 3SP property.*

But still, Ostrovskii actually proves 4.6.c via a result of Beauzamy and Lapresté [20, Thm. 5]: *A Banach space has the alternate Banach-Saks property if and only if no spreading model is isomorphic to l_1.* Let us recall the definition of spreading

model: by an elementary application of Ramsey theory, Brunel and Sucheston (see
[20]) show that any bounded sequence (x_n) in a Banach space admits a subsequence
with the property that for every finite sequence $(\lambda_i)_{1 \leq i \leq n}$ of scalars the limit

$$L(\lambda_1,\ldots,\lambda_n) = \lim_{\substack{\nu \to \infty \\ \nu < n(1) < \ldots < n(k)}} \left\| \sum_{j=1}^{k} \lambda_j x_{n(j)} \right\|$$

exists. It can be proved that L is a norm on the space of finite sequences of scalars
when the starting sequence admits no convergent subsequences. Given thus such a
good sequence in X, the completion of the space of finite sequences with respect to
L is called a *spreading model* of X.

Ostrovskii proves:

Theorem 4.6.d. *The property of not having spreading models isomorphic to l_1 is
a 3SP property.*

Proof. Let $0 \to Y \to X \to Z \to 0$ be a short exact sequence and let (x_n) be a
normalized sequence in X having spreading model δ-isomorphic to l_1; i.e., for every
finite sequence $\lambda_1,\ldots,\lambda_k$ of scalars

$$\delta \sum_{j=1}^{k} |\lambda_j| \leq \lim_{\substack{\nu \to \infty \\ \nu < n(1) < \ldots < n(k)}} \left\| \sum_{j=1}^{k} \lambda_j x_{n(j)} \right\| \leq \delta^{-1} \sum_{j=1}^{k} |\lambda_j|.$$

Since Z has not l_1 as spreading model, and passing to a subsequence if
necessary, there is some finite sequence $\lambda_1,\ldots,\lambda_k$ of scalars such that for all finite
sequences $\{n(i)\}_{1 \leq i \leq k}$ one has

$$\left\| \sum_{j=1}^{k} \lambda_j q\, x_{n(j)} \right\| < \delta/2.$$

If $v_n = \sum_{j=n_k+1}^{n_k+k} \lambda_j x_j$ then there exist points (u_n) in X with $\|u_n\| < \delta/2$ and
$q u_n = q v_n$. The points $y_n = u_n - v_n$ belong to Y and generate a spreading model
isomorphic to l_1 since

$$\frac{\delta}{2} \sum_{i=1}^{m} |\mu_i| \leq \lim_{\substack{\nu \to \infty \\ \nu < n(1) < \ldots < n(k)}} \inf \left\| \sum_{i=1}^{m} \mu_i \sum_{j=1}^{k} \lambda_j x_{n(i)k+j} \right\| - \frac{\delta}{2} \sum_i |\mu_i|$$

$$\leq \lim_{\substack{\nu \to \infty \\ \nu < n(1) < \ldots < n(m)}} \inf \left\| \sum_{i=1}^{m} \mu_i y_{n(i)} \right\|. \qquad \square$$

Question. *Is "not having spreading models isomorphic to l_p" a 3SP property?*

Ostrovskii, answering an e-mail question in [59], remarked to us that

Theorem. 4.6.e. *The p-Banach-Saks and W_p properties and their dual properties are not 3SP.*

Proof. It follows from an examination of Kalton-Peck Z_p spaces. The estimate given in the section 3.2 shows that the sequence $((0,e_n))_n$ is weakly null and admits no subsequences satisfying upper-p-estimates.   □

A partial positive result to the 3SP property for the *p-Banach-Saks* property appears in [303]:

Proposition 4.6.f. *If Y and X/Y have the p-Banach-Saks property then X has the r-Banach-Saks property for all $r<p$.*

Also, Ostrovskii [272] and Rakov [303] prove.

Proposition 4.6.g. *If Y and X/Y have the weak Banach-Saks property and one of them has the Schur property then X has the weak Banach-Saks property.*

4.7   Weak sequential completeness

Infinite dimensional Banach spaces are never weakly complete. They can, however, be weakly sequentially complete (weakly Cauchy sequences are weakly convergent; *wsc*, in short). Reflexive spaces or $L_1(0, 1)$ are *wsc*. As we already said, many authors (Godefroy, Jarchow, Onieva...) proved that being *wsc* is a 3SP property. Let us give yet another proof due to Rakov [302].

Theorem 4.7.a. *Weak sequential completeness is a 3SP property.*

Proof. Assume that X is not *wsc* and that Y is a subspace of X. Let (x_n) be a weak* convergent sequence of X to some element $x^{**} \in X^{**} \backslash X$. Two possibilities exists:

i) for some element $y^{**} \in Y^{**}$ it occurs that $x^{**} + y^{**} \in X$. Then since x^{**} is a Baire-1 element of X^{**}, y^{**} is a Baire-1 element of Y^{**} and Y is not *wsc*.

ii) $(x^{**} + Y^{**}) \cap X = \varnothing$. Since if $q: X \to X/Y$ is the quotient map, (qx_n) is weak* convergent to $q^{**}x^{**} = x^{**} + Y^{**}$ and therefore $q^{**}x^{**} \notin X/Y$. In this way X/Y is not wsc.   □

By the exactness of the functors ^{up} and ^{co} both *super-wsc* and *wsc^{co}* are 3SP properties. No proper description is known of these space ideals (cf. [184]).

4.8 Weak* sequential compactness

Weak sequential compactness of the unit ball is equivalent to reflexivity and, therefore, a $3SP$ property. Having weak* compact dual ball and having weak* sequentially compact dual ball are, however, different properties: the unit ball of $(l_\infty)^*$ is weak* compact but not weak*-sequentially compact. We present now Bourgain's proof that having weak* sequentially compact dual ball is not a $3SP$ property (see also [89, p. 237] and [161]).

Theorem 4.8.a. *To have weak* sequentially compact dual ball is not a $3SP$*

Proof. Let $P_\infty(\mathbb{N})$ be the class of all infinite subsets of $\mathbb{N}$. Let $\{M_\alpha: \alpha < \omega_1\}$ be a well-ordered subset of $P_\infty(\mathbb{N})$ satisfying (see 6.8.f for an explicit construction)

(†)   If $\alpha < \beta$, either $M_\alpha \cap M_\beta$ is finite or $M_\beta \backslash M_\alpha$ is finite;
(††) If $M \subset \mathbb{N}$, there exists some $\alpha \in I$ such that $M \cap M_\alpha$ and $M \backslash M_\alpha$ are both infinite.

Consider the Stone compact K constructed with the algebra generated by this family as in 3.5. The counterexample shall be the sequence $0 \to c_0 \to C(K) \to C(K \backslash \mathbb{N}) \to 0$. It is simpler in this case to see $C(K)$ as the subspace of l_∞ spanned by the set D of characteristic functions $\mathbf{1}_\alpha$ of the elements M_α and of the characteristic functions $\mathbf{1}_n$ of the sets $\{n\}$.

The condition (††) guarantees that (δ_n), the evaluation functionals at h_n has not weak* convergent subsequences: given $Z = \{(n(j)_{j \in \mathbb{N}}\}$ there is some M_α such that both $Z \cap M_\alpha$ and $Z \cap (\mathbb{N} \backslash M_\alpha)$ are infinite. Therefore the sequence $(\delta_{n(j)}(\mathbf{1}_\alpha))_j$ contains an infinity of 1's and 0's.

To prove that $(X/c_0)^*$ has weakly* sequentially compact unit ball observe that if (f_n) is a sequence in the unit ball without weak* convergent subsequences, then the sequence $(f_n|_D)_{n \in \mathbb{N}}$ has not pointwise convergent subsequences. By Rosenthal's dichotomy (see [89, Chapter XI]), there exists a subsequence, say (f_n) itself, and numbers $\delta > 0$ and r, such that the sets $A_n = \{t \in D: f_n(t) > r + \delta\}$ and $B_n = \{t \in D: f_n(t) < r\}$ form an independent sequence; i.e., for any finite disjoint subsets P and Q of $\mathbb{N}$

$$\bigcap_{n \in P} A_n \cap \bigcap_{n \in Q} B_n \neq \emptyset.$$

There is no loss of generality assuming that $r = -\delta/2$ (adding some fixed vector, for instance). The independence of (A_i, B_i) allows one to select, for every $n \in \mathbb{N}$, points $t_1, \dots, t_n$ as follows:

$$t_1 \in \left[\bigcap_{i=1}^{2^{n-1}} A_i \right] \cap \left[\bigcap_{i=2^{n-1}+1}^{2^n} B_i \right],$$

$$t_2 \in \left[\bigcap_{i=1}^{2^{n-2}} A_i \right] \cap \left[\bigcap_{i=2^{n-2}+1}^{2^n} B_i \right] \cap \left[\bigcap_{i=2^{n-1}+1}^{3 \cdot 2^{n-2}} A_i \right] \cap \left[\bigcap_{i=3 \cdot 2^{n-2}+1}^{2^n} B_i \right],$$

...

$$t_n \in A_1 \cap B_2 \cap A_3 \cap B_4 \cap \cdots \cap A_{2^n-1} \cap B_{2^n},$$

so that for every choice of signs $\varepsilon_1, \ldots, \varepsilon_n$ there exist $1 \leq k(1), \ldots, k(n) \leq 2^n$ so that $sign\, f_{k(i)} = \varepsilon_i$. Therefore, if we denote $t_i = 1_{\alpha(i)}$,

$$\left\| \sum_{i=1}^n \lambda_i 1_{\alpha(i)} \right\| \geq \frac{\delta}{2} \sum_{i=1}^n |\lambda_i|.$$

Notice that if k sets $M_{\alpha(1)} \ldots M_{\alpha(k)}$ are pairwise disjoint (except for a finite number of points) then

$$\left\| \sum_{i=1}^{i=k} \lambda_i (\chi_{\alpha(i)} + c_0) \right\| = max\, |\lambda_i|$$

while if they form a chain (except for a finite number of points) $M_{\alpha(1)} \subset M_{\alpha(2)} \subset \ldots \subset M_{\alpha(k)}$ then

$$\left\| \sum_{i=1}^{i=k} \lambda_i (\chi_{\alpha(i)} + c_0) \right\| = \max_j \left| \sum_{i=1}^{i=j} \lambda_i \right|.$$

It is quite clear that a sequence in these conditions cannot be equivalent to the canonical basis of l_1. Finally, observe that if $n \geq k^2 + 1$, a subset of k elements of $\{M_{\alpha(1)} \ldots M_{\alpha(n)}\}$ can be found such that one of the preceding two alternatives holds: this can be easily seen as follows: put $M_\alpha \to M_\beta$ if $M_\alpha \subseteq M_\beta$ and $M_\alpha \downarrow M_\beta$ if $M_\alpha \cap M_\beta = \emptyset$. Now, draw a picture with all the elements: put M_β on the right of M_α if $M_\alpha \to M_\beta$, and put M_β below M_α if $M_\alpha \downarrow M_\beta$. Quite clearly $k^2 + 1$ elements fill a $k \times k$ square.   □

This example is surprising since the fact that the space $C(K \setminus \mathbb{N})$ has dual ball weak* sequentially compact, while $C(K)$ does not, seems to mean that evaluation at points of $\mathbb{N}$ are the only functionals that fail to admit weak* convergent subsequences. Schlumprecht's counterexample for the Gelfand-Phillips property in 6.8 is also a counterexample for the weak* sequential compactness of the dual ball.

4.9   Mazur property or *d*-completeness

A Banach space X is said to have *Mazur's property* (also, *d-complete* [217] or μB-spaces [358]) if every element of X^{**} that is weakly* sequentially continuous is weak* continuous (and thus an element of X). The motivation to consider *d*-complete spaces stems from their applications in measure theory; for instance, in *d*-complete spaces every bounded Dunford-integrable function is Pettis-integrable [107, p.564]. It was proved in [217, Cor 2.5] that *if $X \in Sep^{co}$ then X and X^* are d-complete*. Grothendieck *d*-complete spaces are reflexive. If K is a scattered compact space and there exists a countable ordinal α so that $K^{\alpha} = \varnothing$ then $C(K)$ is *d*-complete. However, $C[0,\omega_1]$ is not *d*-complete [217, Thm 4.1. and Ex. section 3]. Further properties of spaces with Mazur's property can be seen in [237].

Let S_1 be the functor (and "intermediate class") assigning to a Banach space X the normed space

$$S_1(X) = \{f \in X^{**} : f \text{ is weak*-sequentially continuous on } (X^*, w^*) \}.$$

Observe that X *d*-complete means that $X = S_1(X)$. If I denotes the identity functor and $\tau: I \to F_1$ the natural transformation that embeds X into $S_1(X)$, then 2.3.a would conclude that $\{X \in \mathbf{Ban}: \tau X: X \to S_1(X) \text{ is surjective}\}$ is a 3*SP* class *if S_1 were left exact*. Nevertheless, the functor S_1 is not left exact (see the "intermediate classes method at 2.3.b). Therefore we can only show a partial result: if $0 \to Y \to X \to Z \to 0$ is a short exact sequence that S_1 transforms in a left exact sequence and both Y and Z are *d*-complete, X is *d*-complete. Observe that the hypothesis of exactness at X means that if $f \in Y^{**}$ is a weak*-sequentially continuous element of X^{**} then it is a weak*-sequentially continuous element of Y^{**}. This happens when weak* convergent sequences in Y^* can be lifted to weak* convergent sequences in X^*, something that can be done when X has weak* sequentially compact dual ball (2.4.f). This partial answer also appears in [217].

4.10   Weakly compactly generated spaces

A Banach space X is said to be *weakly compactly generated* (WCG) if there exists a weakly compact subset K of X such that *span K* is dense. Separable spaces are WCG spaces; $c_0(I)$-spaces, I any set, are also WCG, since they are generated by the image of the unit ball of $l_2(I) \to c_0(I)$.

First counterexample. It was Lindenstrauss [245, p.243] who first showed that being WCG was not a 3*SP* property. His example was the sequence

$$0 \to C[0,1] \to D[0,1] \to c_0(I) \to 0.$$

The space $D[0,1]$ appears described in the 3.5 and in 5.5.b. It is not WCG because every weakly compact subset of $D[0,1]$ is separable (as it happens in every space whose dual is weak* separable - in $D[0,1]$ evaluation functionals at rational points form a weak* total set in $D[0,1]^*$) while the space is not.

A second counterexample is provided by the James Tree space JT (see 4.14.e) and the sequence

$$0 \to B \to JT^* \to l_2(\Gamma) \to 0$$

where B is the predual of JT: the space B is WCG since it is separable. The space JT has all its even duals WCG while its odd duals are not WCG. This even shows that X^{**} need not be WCG although both X and X^{**}/X are WCG.

Problem. Let us remind that it is still an open problem to know if when X^{**} is WCG then X is WCG.

The third and basic counterexample is due to Johnson and Lindenstrauss [193].

Theorem 4.10.a. *To be WCG is not a 3SP property.*

Proof. Consider an *almost disjoint* uncountable family $\{A_i\}_{i \in I}$ of infinite subsets of $\mathbb{N}$, i.e., a family such that the intersection of any two different members of the family is finite: it can be obtained enumerating the rationals and taking, for each irrational, the indices corresponding to a sequence of rationals converging to it.

The Johnson-Lindenstrauss space JL is defined as the completion of the linear span of $c_0 \cup \{1_i : i \in I\}$ in l_∞ (here 1_i denotes the characteristic function of A_i) with respect to the norm:

$$\| y = \textstyle\sum_{j=1}^{j=k} a_{i(j)} 1_{i(j)} + x \| = max \{ \|y\|_\infty, \ \|(a_i)_{i \in I}\|_{l_2(I)} \} .$$

This norm is well defined since

$$a_{i(j)} = lim_{n \in N_{i(j)}} y_n .$$

Obviously c_0 is a subspace of JL, and let us verify that JL/c_0 is isomorphic to $l_2(I)$: if $y \in V$, then

$$\|y\| \geq \|(a_{i(j)})\|_2.$$

On the other hand, given a finite sequence $(a_{i(j)})$, consider the different sets $A_{i(j)}$: they intersect in a finite set, and therefore it is possible to find a (finite) sequence $x \in c_0$ that anihilates that common part; this transforms the l_2-norm into the *sup* norm and thus

$$\left\| \sum_{j=1}^{k} a_{i(j)} 1_{i(j)} + x \right\| \leq \left\| (a_{i_{i \in I}}) \right\|_{l_2(I)}.$$

It only remains to prove that JL is not WCG. This follows from the facts that JL is not separable and JL^* admits a countable weak* total set: observe that the elements of JL are bounded sequences endowed with a norm finer than the l_∞ norm; thus, the coordinate functional are continuous and form a weak* total set in JL^*, which implies that every weakly compact set of JL is separable. □

A fourth counterexample using $C(K)$ spaces. It is simple to determine when a $C(K)$ space is WCG (see [8, Prop. 1] or also [82, lemma 7.4]).

Lemma 4.10.b. *The space $C(K)$ is WCG if and only if K is homeomorphic to a weakly compact set of a Banach space (an Eberlein compact).*

We shall only need the "only if" implication:

Observe that if W is a weakly compact subset of a Banach space X then the restriction map $R: X^* \to C(W)$ is weak*-to-weak continuous; and this is so because it is obviously Mackey(X^*,X)-to-norm continuous and so continuous with respect to their respective weak topologies (see [225]). From this, it immediately follows:

Proposition 4.10.c. *A Banach space X is WCG if and only if there exists a Banach space M and an injective weak*-to-weak continuous operator $T: X^* \to M$.*

Because for one implication (only if) take $M = C(W)$ together with the preceding argument; and for the other (if) note that T^* has dense range and is weak*-to-weak continuous.

End of the proof of the lemma. If $C(K)$ is WCG, for some space M there is an injective weak*-to-weak continuous operator $T: C(K)^* \to M$; this makes K (in fact, its homeomorphic image in $C(K)^*$) homeomorphic to a weakly compact set: its image by T. □

Let us then show that there is a compact space $K\Delta$ that is not an Eberlein compact for which there exists a short exact sequence

$$0 \to c_0 \to C(K\Delta) \to c_0(I) \to 0.$$

The compact space $K\Delta$ is the Stone compact generated by an uncountable almost disjoint family of infinite subsets of $\mathbb{N}$. To obtain this family it is useful to think of $\mathbb{N}$ as the nodes of a dyadic tree, and the elements of the family are the branches of the tree. In the Stone compact $K\Delta$ there are three classes of points: the points of $\mathbb{N}$ (Boolean functions h such that $h(\{n\})=1$ for some $n \in \mathbb{N}$), the branches (Boolean functions h such that $h(r_\alpha)=1$ for some branch r_α), and the point ∞ (Boolean function h_∞ such that $h_\infty(r_\alpha)=0$ for all branches r_α and $h_\infty(\{n\})=0$ for all $n \in \mathbb{N}$). The derived set $(K\Delta)'$ is the set of branches and ∞; the second derived set $(K\Delta)''$ is $\{\infty\}$ and the third derived set is empty. It is clear that $C(K\Delta)/c_0 = C(K\Delta\backslash\mathbb{N}) = C(\{\text{branches}\}\cup\{\infty\}) = c(I)$ since the branches are mutually isolated and ∞ is their only accumulation point. It only remains to prove one thing.

Claim. $K\Delta$ is not an Eberlein compact.

Proof 1. Recall that separable weakly compact sets are metrizable. Since the set of branches is uncountable and they are mutually isolated points, $K\Delta$ is not metrizable (the point ∞ cannot have a countable fundamental system of neighborhoods). Since $K\Delta$ is separable ($\mathbb{N}$ is dense) it cannot be homeomorphic to a weakly compact set.

Proof 2. One could also directly prove that c_0 is not complemented in $C(K\Delta)$; and then appeal to a fundamental result of Amir and Lindenstrauss [8]:

Proposition 4.10.d. *In a WCG space every separable subspace is contained into a complemented separable subspace.*

and to Sobczyk's theorem (c_0 is complemented in any separable space containing it). In fact, one does not need the full strength of Amir-Lindenstrauss result: it is enough to take into account Rosenthal's observation (see [354]) that Veech's proof of Sobczyck's theorem just requires that the weak*-closure of a sequence in $Ball(X^*)$ is metrizable, something that happens in *WCG* spaces (in fact, the dual ball of a *WCG* space is, in its weak* topology, an Eberlein compact; see [87, p. 148]).

To prove that c_0 is not complemented in $C(K\Delta)$ one can proceed very much as in the proofs that c_0 is not complemented in l_∞. A possibility is to prove that the evaluation functionals on the points of $\mathbb{N}$ do not admit a pointwise-zero extension to all of $C(K\Delta)$. This is the way Moltó chooses in [261], where $K\Delta$ is presented under homeomorphic clothes as a Rosenthal compact (see also [82, p.260]). Another possibility is to realize (see [178]) that the kernel of a continuous functional $\phi \in C(K\Delta)^*$ that vanishes on c_0 must contain an uncountable quantity of characteristic functions of elements of the almost-disjoint family. The range of a projection onto c_0 would be, at the same time, c_0 and a countable intersection of kernels of such functionals, which is impossible. $\square$

A question. The Johnson-Lindenstrauss space shows that there exist nontrivial quasilinear maps $l_2(\Gamma) \to c_0$. It is unknown, however, if there exist nontrivial pseudo-linear maps

$$\Omega: l_2(\Gamma) \to c_0.$$

It could be a good warning to observe that although nontrivial quasilinear maps $c_0(\Gamma) \to c_0$ exist, as the space $C(K\Delta)$ shows, nontrivial pseudolinear maps $\Omega: c_0(\Gamma) \to c_0$ do not exist since $Ball\,(c_0(\Gamma)^*) = Ball(l_1(\Gamma))$ is irreducible [362].

Some positive results can be obtained when Y is reflexive or X/Y is separable. Observe that the counterexamples have been obtained exchanging these two conditions: Y separable and X/Y reflexive.

Proposition 4.10.e. *Let $0 \to Y \to X \to Z \to 0$ be a short exact sequence. If Y is reflexive and Z is WCG, then X is WCG.*

Proof. Let K a weakly compact set in Z spanning a dense subspace. Take a bounded subset A of X such that $q(A)=K$. Since Y is reflexive, using the lifting 2.4.b one obtains that the set A is relatively weakly compact. The relatively weakly compact set $A \cup Ball(Y)$ spans a dense subspace of X. $\square$

Observe that the lifting of weakly compact sets is not always possible: weakly compact sets in l_∞ are separable while l_∞/c_0 admits nonseparable weakly compact sets (the canonical image of the unit ball of $l_2(\Gamma)=JL/c_0$), and these cannot be lifted.

Proposition 4.10.f. *Let Y be a subspace of X such that X/Y is separable. Then X is WCG if and only if Y is WCG.*

Proof. Assume that Y is WCG. Take a norm null sequence (x_n) in X such that (qx_n) spans a dense subspace of X/Y. If K is a weakly compact set in Y spanning a dense subspace, $K \cup \{x_n\}_n$ is relatively weakly compact and spans a dense subspace in X.

Assume now that X is WCG and find a set $\{x_\alpha: \alpha \in \Lambda\}$ in X that is weakly homeomorphic to the one-point compactification of a discrete set that spans a dense subspace [245, p. 249]. Since X/Y is separable there is some $\Lambda_0 \subset \Lambda$ with $\Lambda \backslash \Lambda_0$ countable such that $x_\alpha \in Y$ for all $\alpha \in \Lambda_0$. The subspace M spanned by $\{x_\alpha : \alpha \in \Lambda_0\}$ in Y is WCG and Y/M is separable, and thus the first part proves that Y is WCG. $\square$

Still another result showing that the Johnson-Lindenstrauss example is "sharp":

Proposition 4.10.g. *If Y^{**} is separable and X/Y is reflexive then X and X^{**} are WCG.*

Proof. Since $Y^\perp =(X/Y)^*$ is reflexive and Y^* separable, applying 4.10.f to the dual sequence $0 \to (X/Y)^* \to X^* \to Y^* \to 0$ one sees that X^* is WCG. Thus, if F is a separable subspace of X^* such that $Y^\perp + F = X^*$, by 4.10.d there is no loss of generality assuming that F is complemented; i.e., $X^*=F \oplus G$. Since

$$G = Y^{\perp} / (Y^{\perp} \cap F),$$

G is reflexive and, in particular, $G = (^{\perp}G)^{\perp}$. Set $H = {}^{\perp}G$.

Since $H^* = F$, H is separable and it follows from the hypothesis that $H^{\perp\perp}$ is separable. Thus, $X^{**} = F^{\perp} \oplus H^{\perp\perp}$ with $F^{\perp}$ reflexive. Hence X^{**} is WCG. The hypotheses imply that $X^{**}/X = Y^{**}/Y$ is separable; so, applying 4.10.e to X and X^{**} we conclude that X is WCG. $\qquad\square$

Although JL is not WCG, its dual space JL^* is isomorphic to $l_1 \oplus l_2(\Gamma)$ and thus it is WCG.

Question. *Is WCG^{dual} a 3SP property?*

A good try to this seemingly hard question could be to obtain a counterexample for the 3SP property of WCG starting with a dual space since a counterexample to WCG^{dual} is equivalent to the existence of a sequence $0 \to Y^* \to X \to R \to 0$ with R reflexive and X not WCG. In [63] it is shown that if $0 \to Y^* \to X \to Z \to 0$ is an exact sequence where Y^* is a separable dual and Z is WCG then also X is WCG.

Theorem 4.10.h. *WCG^{co} is not a 3SP property.*

Proof. The proof depends on a counterexample for WCG to be constructed in 4.14; it has the form $0 \to X \to X^{**} \to X^{**}/X \to 0$ with X separable. For separable spaces there always exist a space M such that $M^{**}/M = X$, so the observation 2.2.k applies. $\qquad\square$

An intermission: proof of a theorem of Valdivia. Let us give

a short proof of the following result of Valdivia [349] that shall be essential for what follows.

Proposition 4.10.i. *Separable ^{co} = Reflexive $\oplus$ Separable $^{dual\ dual}$*

Proof. One implication is obvious, so let X be a Banach space such that X^{**}/X is separable. Let us consider the following closed subspace of X^{**} containing X:

$$G = \left\{ \bigcup M^{\perp\perp} : M \subset X,\ M\ separable \right\}.$$

Since G/X is separable, there is a sequence (x_n) of elements of G such that $X \cup \{x_n\}_{n \in \mathbb{N}}$ is dense in G. For every $n \in \mathbb{N}$ let us select a separable subspace M_n of X such that $x_n \in M_n^{\perp\perp}$. Let us denote by N the separable subspace of X spanned by the union of all M_n. It was proved in 2.10.k that when N is a closed subspace of X then $X + N^{\perp\perp}$ is closed in X^{**}, and thus we conclude that $X + N^{\perp\perp} = G$. Observe that weak* cluster points of sequences of elements of X belong to

$X+N^{\perp\perp}$. Hence X/N is reflexive. Moreover, N^{**}/N is separable because it is a subspace of X^{**}/X. Thus, also N^{**} is separable, and it follows from 4.10.g that X is *WCG*. Accordingly, there is some separable complemented subspace V containing N that obviously has X/V reflexive. $\square$

4.11   Asplund spaces

A Banach space is said to be an *Asplund space* if every separable subspace has separable dual. The topological version of this last formulation could be: every subspace has the same density character as its dual. Subspaces of Asplund spaces are obviously Asplund spaces. Asplund spaces coincide with spaces having duals with the Radon-Nikodym property [330]. Since the *RNP* is a 3*SP* property, the same occurs with the RNP^{dual}. Therefore,

Theorem 4.11.a. *To be an Asplund space is a 3SP property.*

The first proof of this result appeared in [265]. Nevertheless, a more direct proof can be shown following [363].

2nd *Proof.* Let $0 \to Y \to X \to Z \to 0$ be a short exact sequence in which Y and Z are Asplund spaces. Let S be any separable subspace of X. Since $\overline{Y+S}/Y$ is separable, it is possible to choose some separable subspace Z of X such that $Y+Z = \overline{Y+S}$. It can be easily assumed that Z contains S (replace Z by $\overline{Z+S}$ if necessary). Since $Y+Z$ is closed $Z/(Z\cap Y) \cong (Y+Z)/Y \subset X/Y$, which yields that $Z/(Z\cap Y)$ has separable dual. Since $Z\cap Y \subset Y$ also has a separable dual then, by the 3*SP* property of separability, Z^* is separable. Hence, S^* is separable. $\square$

Recall a well-known result [93, p. 87]:

Proposition 4.11.b. WCG^{dual} *spaces are Asplund spaces.*

Proof. In fact, quotients of *WCG* spaces are *WCG*. So, if X^* is *WCG*, and Y is a separable subspace of X, then Y^* is *WCG*. In every weakly compact subset $K \subset Y^*$ the weak* and weak topologies coincide. But Y separable implies K compact metric space. Therefore K is weak* separable; hence weakly separable. This fact and the Hahn-Banach theorem imply that K is norm separable. It follows that separable subspaces of X have separable dual. $\square$

4.12   The Plichko-Valdivia property

Let us say that a Banach space X has the *Plichko-Valdivia* property (*PV*, in short) if it can be decomposed in the form $X = S \oplus R$, where S is separable and R is reflexive; i.e.,

$$PV = Separable \oplus Reflexive.$$

It is clear that the *PV* property implies *WCG* and that separable and reflexive spaces have *PV*; moreover.

Proposition 4.12.a. *PV = Reflexive-by-Separable*.

Proof.   Let $0 \to Y \to X \to Z \to 0$ be an exact sequence with Y reflexive and Z separable. By lemma 2.9.a, there is a separable subspace N of X such that $Y+N=X$. Since *PV* implies *WCG*, the separable subspace N is contained in a separable complemented subspace M, whose complement must be reflexive.   □

Theorem 4.12.b. *PV is not a 3SP property*.

Proof.   The sequence $0 \to c_0 \to JL \to l_2(I) \to 0$ together with 4.10.a prove the result.   □

The space $JL^* = l_1 \oplus l_2(I)$ is an example of space with the *PV* property such that its predual *JL* does not have it. Therefore *JL* has the PV^{dual} property. The following results are in [63].

Proposition 4.12.c. $PV^{dual} = Separable^{dual}$-by-Reflexive.

Proof. If $X \in PV^{dual}$ then $X^*=R \oplus S$ where R is reflexive and S separable. Clearly $R=(^\perp R)^\perp$ and thus $X/^\perp R$ is reflexive since $(X/^\perp R)^* = (^\perp R)^\perp = R$ is reflexive; and $(^\perp R)^* = X^*/(^\perp R)^\perp = X^*/R = S$ is separable. Conversely, assume that there is an exact sequence $0 \to M \to X \to X/M \to 0$ where M has separable dual and X/M is reflexive. The dual sequence $0 \to (X/M)^* \to X^* \to M^* \to 0$ is also exact. Since $X^*/M^\perp = M^*$ is separable, there is a separable subspace F of X^* such that $X^*=M^\perp+F$ with $M^\perp$ reflexive. Thus, X^* is *WCG* and F is contained in a complemented separable subspace G of X^*. Therefore, $X^* = G \oplus H$ where H is reflexive because it is isomorphic to a subspace of $M^\perp =(X/M)^*$.   □

It is clear that PV^{dual} implies *Asplund*: *PV* implies *WCG* plus 4.11.b.

Proposition 4.12.d. $PV^{dual} = (Separable$-by-Reflexive$) \cap$ *Asplund*.

Proof. It follows from 4.12.c that PV^{dual} implies both *Separable-by-Reflexive* and Asplundness. It is obvious that an Asplund space that is *Separable-by-Reflexive* is also *Separable^{dual}-by-Reflexive*.   □

Theorem 4.12.e. PV^{dual} *is a 3SP property.*

Proof. Since both *Separable-by-Reflexive* and *Asplund* are 3SP properties (see 2.9 and 4.11.a).   □

Nevertheless, *Separable^{dual} ⊕ Reflexive* is not a 3SP property as the sequence 0 → c_0 → *JL* → $l_2(I)$ → 0 shows. The space *JL* belongs to *Separable^{dual}-by-Reflexive* but not to *Separable^{dual} ⊕ Reflexive*. Also, observe that *Separable^{dual} ⊕ Reflexive* = *Reflexive-by-Separable^{dual}* since this last property implies, simultaneously, (*Separable ⊕ Reflexive*) and *Asplund*.

As a consequence of results already established:

Proposition 4.12.f. $PV^{dual\ dual}$ *and Asplund = Separable ^{co}.*

Observe that *Asplund* cannot be deleted here; the James-Tree space *JT* (see 4.14.e) is a separable dual space such that $JT^{**}=JT \oplus l_2(\Gamma)$; thus, it is $PV^{dual\ dual}$ but not Asplund. The same reasoning as in 4.10.h gives:

Theorem 4.12.g. PV^{co} *is not a 3SP property.*

A related property has been defined by Vasak [352].

Separately distinguished spaces. Property *Reflexive ⊕ Separable ^{dual}* has been called by Vasak *separately distinguished* (SD). Clearly *Cech^{dual} ⇒ SD*, and it is equally clear that *SD* implies *PV*.

Theorem 4.12.h. *Property SD is not a 3SP property.*

Proof. Since *SD* implies *WCG*, the sequence 0 → c_0 → *JL* → $l_2(\Gamma)$ → 0 gives a counterexample.   □

4.13   Polish spaces

A Banach space X is said to have a *Polish* ball if $(B_X,$ *weak*) is homeomorphic to a complete metrizable separable space. Observe that completeness is not preserved by homeomorphisms, reason for which spaces with Polish ball need not be reflexive. Clearly, the metrizability assumption implies that X^* is separable. If, conversely, X^* is separable, there is a metric that induces the weak topology on B_X, but this metric is not complete unless X is reflexive (in fact, the completion of $(B_X,$ *weak*) is $(B_{X^{**}},$ *weak**)). For instance, c_0 has not Polish ball in the weak topology:

its unit ball can be written as

$$Ball(c_0) = \cup_n A_n = \{ x \in Ball(c_0) : |x(k)| \leq 1/2 , k \geq n \}$$

where the sets A_n are weakly closed with empty weak interior; this means that $(Ball(c_0)$, weak) is not a Baire space and, a fortiori, it is not a complete metric space. The following result is of Edgar and Wheeler [108].

Proposition 4.13.a. *To have Polish ball is equivalent to have separable dual and the PCP.*

We shall see in 4.17 that *PCP* is a 3*SP* property, and thus

Theorem 4.13.b. *To have Polish ball is a 3SP property.*

We shall denote by *Polish* the class of spaces having Polish ball. It is obvious that *Polish* implies *PV*.

4.14   Cech completeness

Recall from general topology that a topological space (E,τ) is Cech complete if it admits a complete sequence $(\mathbf{C}_n)$ of open covers; complete means that every family $\mathscr{F}$ of closed sets of (E,τ) with the finite intersection property *and* small with respect to $(\mathbf{C}_n)$ - in the sense that given $\mathbf{C}_n$ and $F \in \mathscr{F}$ there is $U \in \mathbf{C}_n$ containing F - has nonempty intersection. Metric complete spaces are Cech complete as can be seen taking as $\mathbf{C}_n$ the open sets of diameter less than $1/n$ and recalling Cantor's intersection argument. This yields that a Polish space is homeomorphic to a Cech complete space. Let us call a Banach space X *Cech complete* if $(B_X$, weak) is Cech complete. The following result from general topology clarifies the structure of Cech complete spaces.

Lemma 4.14.a. *A topological space (E,τ) is Cech complete if and only if (E,τ) is a G_δ set in some (any) compactification.*

Therefore, $(B_X$, weak) is Cech complete if and only if B_X is a G_δ of $(B_{X^{**}}$, weak*); and this happens if and only if X is a G_δ in $(X^{**}$, weak*); if and only if $B_{X^{**}} \backslash B_X$ is a K_σ in $(B_{X^{**}}$, weak*); and if and only if $X^{**} \backslash X$ is a K_σ in $(X^{**}$, weak*). Moreover, observe that a linear isomorphism preserves the G_δ-character of X into $(X^{**}$, weak*). These equivalent formulations and several result of general topology allow one to prove:

Lemma 4.14.b. *If X is a separable Banach space with Cech complete unit ball then X^* is separable.*

From where it can be deduced [108, Corol. 3.2]:

Proposition 4.14.c. *A Banach space has Polish ball if and only if it has Cech complete ball and is separable.*

We shall denote *Cech* the class of Banach spaces having Cech complete ball. An interesting characterization of Cech complete Banach spaces is [108, Thm. B]:

Proposition 4.14.d. *Cech = Reflexive $\oplus$ Polish.*

Proof. If $(B_X, weak)$ is Cech complete then X is a G_δ in X^{**} and therefore $X = \bigcap_n U_n$ where U_n are weak* open neighborhoods of 0 in X^{**}. Let F_n be a finite set of X^* such that U_n contains the basic weak*-neighborhood that F_n determines. Let $G = \overline{span} \bigcup F_n$ in X^*. Since $G^\perp$ is weak*-closed in X^{**} and $G^\perp \subset \bigcap_n U_n = X$, it turns out that the unit ball of $G^\perp$ is weak* = weak compact; i.e., $G^\perp$ is a reflexive subspace of X.

On the other hand, $(X/G^\perp)^* = G$ which implies that $X/G^\perp$ has separable dual. It is therefore possible to find a separable closed subspace S of X such that $X = S + G^\perp$. This yields that X is WCG and thus there is a separable subspace S_1 containing S and complemented in X by a reflexive subspace (a quotient of $G^\perp$). The separable subspace S_1 has Cech complete ball (as a subspace of X) and thus it has a Polish ball. $\square$

Theorem 4.14.e. *Cech completeness is not a 3SP property.*

Proof. We shall consider the exact sequence

$$0 \to B \to JT^* \to l_2(\Gamma) \to 0$$

where B is the predual of the *James-Tree* space. Let us give a description of this space and some of the basic properties required.

The James Tree space. The James-Tree space *JT* appears as a negative answer to the question: does every separable space with nonseparable dual contain l_1? The space *JT* is a dual separable space (it has a boundedly complete basis) that is l_2-saturated and has nonseparable dual. It is defined as the completion of the space of finite sequences over the dyadic tree Δ with respect to the norm

$$\|x\| = sup_{N \in \mathbb{N}} \, sup_{S_1,\ldots,S_N} \left[\sum_{i=1}^{N} \left(\sum_{\alpha \in S_i} x_\alpha \right)^2 \right]^{1/2}$$

where the *sup* is taken over all finite sets of pairwise disjoint segments of Δ. If Δ is identified with $\mathbb{N}$ in some obvious manner, it is not difficult to verify that the vectors $(e_n)_{n \in \mathbb{N}}$ form a boundedly complete basis. Hence, JT is a dual space, $JT=B^*$ where $B=[e_n{}^*]$ in JT^*. That JT is l_2-saturated needs some more work. It also needs some work to prove that if Γ denotes the uncountable set of branches of Δ then the application $L: JT^* \to l_2(\Gamma)$ given by

$$L(f)(r) = \lim_{j \in r} f(e_j)$$

is a well-defined quotient map with kernel B (both results can be seen in [315] or [105]). The nonseparability of JT^* follows from this.

Proof of 4.14.e. Obviously, the quotient $JT^*/B=l_2(\Gamma)$ has a Cech complete ball. The space JT^*, however, cannot have Cech complete ball since it is not *WCG* (it is a nonseparable dual of a separable space). It only remains to prove:

Claim. The space B has Cech complete ball.

Proof of the claim. Let us denote by F_n the nth-"*floor*" of the dyadic tree, i.e., the set $F_n = \{ 2^n, \dots, 2^{n+1}-1 \}$; and let $M_n: JT^* \to \mathbb{R}$ be the function

$$M_n(f) = \left(\sum_{i \in F_n} |f(e_i)|^2 \right)^{1/2}$$

that is obviously weak*-continuous. The exactness of the sequence

$$0 \to B \to JT^* \to l_2(\Gamma) \to 0$$

where the quotient map is L implies that

$$B = \{ f \in JT^* : \lim_{n \to \infty} M_n f = 0 \}$$

The last equality can be written as

$$B = \bigcap_{k=1}^{\infty} \bigcup_{n=k}^{\infty} \left\{ f \in JT^* : M_n(f) < \frac{1}{k} \right\}.$$

The sets $\{ f \in JT^* : M_n(f) < k^{-1} \}$ are weak*-open and thus B has Cech complete ball since it is a G_δ in $(JT^*, weak^*)$. $\qquad\square$

Edgar and Wheeler [108, Prop.4.2.] give a partial positive result.

Proposition 4.14.f. *Reflexive-by-Cech implies Cech.*

If one considers instead spaces having Cech complete dual ball, the following structural decomposition is possible.

Proposition 4.14.g. $Cech^{dual}$ = $Reflexive \oplus Separable^{dual\ dual}$ = $Separable^{co}$

Proof. Recall the second equality was proved in 4.10.i. Now, if X^* has Cech complete ball then $X^{***}\backslash X^*$ is a K_σ in $(X^{***}, weak^*)$. Since $X^\perp$ is weak*-closed, $X^\perp \cap X^{***}\backslash X^* = X^\perp\backslash\{0\}$ is again a K_σ in $(X^{***}, weak^*)$. Thus, $\{0\}$ is a *weak* G_δ in $X^\perp = (X^{**}/X)^*$. If $\{0\} = \cap_{n \in \mathbf{N}} O_n$ where O_n is a weak* open neighborhood of 0 defined by the finite set F_n of elements of X^{**}/X, then $C = \cup_{n \in \mathbf{N}} F_n$ is a countable subset of X^{**}/X such that $C^\perp = \{0\}$. Therefore, the span of C is dense in X^{**}/X.

Conversely, if $X = R \oplus S$ with R reflexive and S^{**} separable, then S^* is a separable dual and must have the *RNP*, and thus the *PCP*. By 4.14.c, S^* has Cech complete ball and the same happens to $X^* = R^* \oplus S^*$ □

Since separability (hence *Separable*^{co}) is a 3*SP* property, one has.

Theorem 4.14.h. *To have Cech complete dual ball is a 3SP property*

The general schema in 2.2.k, the existence of the counterexample $0 \to B \to JT^*$ $\to l_2(\Gamma) \to 0$ and the fact that B is separable yield.

Theorem 4.14.i. $Cech^{co}$ *is not a 3SP property.*

4.15 Between *WCG* and *PRI*

Now we describe properties that can be considered as weak forms of being *WCG*.

***X* is a weak* $K_{\sigma\delta}$ in *X^{**}*.** This property is weaker than *WCG* as can be easily seen: if X is *WCG* then for some weakly compact set W and every $m \in \mathbf{N}$, $X = \cup_n nW + m^{-1}B_X$, which yields

$$X = \bigcap_m \bigcup_n (nW + m^{-1}B_{X^{**}}).$$

As we shall see in 4.15.c, this is not a 3*SP* property.

Weakly K-analytic spaces. A topological space T is said to be *K-analytic* if it is the continuous image of a $K_{\sigma\delta}$ of a compact space. A Banach space X is said to be *weakly K-analytic* (after [338]) if it is a K-analytic space in its weak topology. Banach spaces that are $K_{\sigma\delta}$ in $(X^{**}\ w^*)$ are weakly K-analytic. It is unknown if the converse is true. Compact spaces K for which $C(K)$ is weakly-K-analytic are called *quasi-Eberlein* compacts. Even he following apparently simpler question is open:

if K is a quasi-Eberlein compact, must be $C(K)$ a $K_{\sigma\delta}$ in $C(K)^{**}$?

As we shall see in 4.15.c, weak K-analyticity is not a $3SP$ property.

Weakly countably determined spaces.
A topological subspace $S \subset T$ is said to be *countably determined in* S if there are compacts $(A_i)_{i \in \mathbf{N}}$ in T such that for any x there is a part $A \subset \mathbf{N}$ with $x \in \bigcap \{A_i : i \in A\} \subset S$. A Banach space X is said to be *weakly countably determined* (*WCD*) if it is countably determined in (X^{**}, w^*).

Proposition 4.15.a. *Weakly K-analytic spaces are weakly countably determined.*

As we see in 4.15.c, *WCD* is not a $3SP$ property

Projectional Resolution of the identity.
Let X be a Banach space and let ω_d the smallest ordinal having cardinality equal to the density character of X. A projectional resolution of the identity on X is an ordered family of projections $\{P_\alpha : \omega \leq \alpha \leq \omega_d\}$ satisfying for every α the following conditions:

i) $\| P_\alpha \| = 1$
ii) $P_\alpha P_\beta = P_\beta P_\alpha = P_{min\{\alpha,\beta\}}$
iii) $dens(P_\alpha(X)) \leq card\ \alpha$
iv) $\cup_{\beta < \alpha} P_{\beta+1}(X)$ is norm dense in $P_\alpha(X)$
v) $P_{\omega_d} = id_X$.

The next result was proved by Vasak [352, Thm. 1].

Proposition 4.15.b. *Weakly countably determined spaces admit a PRI.*

Moreover, Fabian proves in [117] that "sometimes" (i.e., when the space admits an equivalent Gateaux differentiable norm, is an Asplund space and all complemented subspaces admit *PRI*) *PRI* implies *WCD*.

The proof.
We establish now the non $3SP$ character of the properties considered so far.

Theorem 4.15.c. *The properties of being 1) $K_{\sigma\delta}$ in X^{**}; 2) weakly K analytic; 3) weakly countably determined; 4) admitting PRI (under some equivalent norm); are not 3SP properties.*

Proof. The short exact sequence (see 4.10)

$$0 \rightarrow c_0 \rightarrow JL \rightarrow l_2(\Gamma) \rightarrow 0$$

gives a counterexample for those properties since c_0 and $l_2(I)$ have all of them while

JL admits none: this is the last result needed (it is due to Plichko [295] and also appears in [118])

Claim. The Johnson-Lindenstrauss space JL does not admit a PRI under any equivalent renorming.

Proof. If *JL* admits some renorming with a *PRI* then (eventually passing to some other *PRI*) some subspace *X* containing the natural copy of c_0 is the range of a projection P_α, $|\alpha| = dens\ X < dens\ JL$ (see [295]). The complement of *X* must be a subspace of $l_2(\Gamma)$, and therefore *JL* can be written as the direct sum of *X* and a Hilbert space; and since weakly compact subsets of *JL* are separable, that Hilbert space is separable, hence $JL = X \oplus l_2$ and $|\alpha| = dens\ X = dens\ JL > |\alpha|$; a contradiction. □

4.16   σ-fragmentability

A topological space (E, τ) is σ-fragmented by a metric ρ in *E* if for each $\varepsilon > 0$ it is possible to write *E* as a countable union of sets with the property that each nonempty subset has a nonempty relatively open subset of ρ-diameter less than ε. This concept has been studied by Jayne, Namioka and Rogers (see [187, 188]). Interesting situations to apply the theory appear when *E* is a Banach space endowed with its weak topology, or a dual space with its weak* topology; or a $C(K)$-space with the topology of pointwise convergence. For instance, in [188, Thm. 6.2] it is proved that a Banach space is Asplund if and only if $(X^*, weak^*)$ is σ-fragmented by the norm.

A Banach space *X* is said to be *σ-fragmentable* if $(X, weak)$ is σ-fragmented by the norm. Spaces admitting a Kadec norm (see 5.20) or weakly-*K*-analytic spaces are σ-fragmentable [188, Corol. 6.3.1 and 6.3.2], while l_∞ is not σ-fragmentable. Ribarska [305] proved that σ-fragmentability is a 3*SP* property.

Theorem 4.16.a. *σ-fragmentability is a 3SP property.*

Proof. Assume that *Y* is a subspace of *X* such that *Y* and $Z = X/Y$ are σ-fragmentable. Let $q: X \to Z$ denote as usual the quotient map and let $B: Z \to X$ be a continuous Bartle-Graves selection for *q*. Let $P: X \to Y$ be the map $P(x) = x - Bqx$. Let $\varepsilon > 0$ fixed.

The norm continuity of *B* yields that for every positive integer $j \in \mathbb{N}$ and every $z \in Z$ there exists $\eta^j(z) > 0$ such that $\|z - z'\| < \eta^j(z)$ implies $\|Bz - Bz'\| < 1/j$. Therefore, for $j, k \in \mathbb{N}$ denote

$$F_k^{(j)} = \{z \in Z : \eta^j(z) > \frac{1}{k}\}.$$

One has that for every $j \in \mathbb{N}$

$$Z = \bigcup_{k=1}^{\infty} F_k^{(j)}.$$

Moreover, if $k \in \mathbb{N}$ is also fixed, the σ-fragmentability of Z allows us to write

$$Z = \bigcup_{m=1}^{\infty} G_m^{(k)}$$

in such a way that for every nonempty $S \subset G_m^{(k)}$ there exists a weakly open subset U of Z such that $U \cap S \neq \emptyset$ and $diam(U \cap S) < 1/k$.

Let $\emptyset \neq S \subset G_m^{(k)} \cap F_k^{(j)}$. Since $S \subset G_m^{(k)}$, there exists a weakly open subset U of Z so that $U \cap S \neq \emptyset$ and $diam(U \cap S) < 1/k$, and given z, z' in $U \cap S$ one has $\| z - z' \| < 1/k < \eta^j(z)$ because $z \in S \subset F_k^{(j)}$; hence $\| Bz - Bz' \| < 1/j$. Now, if l: $\mathbb{N} \times \mathbb{N} \to \mathbb{N}$ is a bijection and we write $l = l(m, k)$ and

$$F_l^j = G_m^{(k)} \cap F_k^{(j)}$$

then, for every $j \in \mathbb{N}$, one has

$$Z = \bigcup_{l=1}^{\infty} F_l^j$$

so that for every $l \in \mathbb{N}$ and every nonempty set $S \subset F_l^j$ there exists a weakly open subset U of Z such that $U \cap S \neq \emptyset$ and $diam\, B(U \cap S) < 1/j$. The σ-fragmentability of the subspace Y implies the existence of a countable cover

$$Y = \bigcup_{n=1}^{\infty} H_n$$

in such a way that for every $n \in \mathbb{N}$ and every nonempty subset S of H_n there exists a weakly open subset V of Y such that $V \cap S \neq \emptyset$ and $diam(V \cap S) < \varepsilon$. We fix $n \in \mathbb{N}$ and construct inductively a family of subsets

$$\{ H_n^\xi : 0 \leq \xi < \xi_n \}$$

of $(H_n, \, weak)$ so that $H_n^0 = \emptyset$, each H_n^ξ is a relatively open subset of

$$H_n \backslash (\bigcup_{\beta < \xi} H_n^\beta),$$

and

$$H_n = \bigcup_{\xi < \xi_n} H_n^\xi.$$

Let us select a nonempty relatively weakly open subset $V \cap H_n$ of H_n having diameter less than ε; here

$$V = \bigcap_{s=1}^{k} \{ y \in Y : h_s(y) > \alpha_s \},$$

with α_s real numbers and h_s norm one elements of X^*, $s = 1, \ldots, k$. For every $y_{n1} \in V \cap H_n$ the number

$$\alpha^{n1} = \min \{ h_s(y_{n1}) - \alpha_s : s=1,\ldots, k \}$$

is positive; hence the set

$$\tilde{V} = \bigcap_{s=1}^{k} \{ y \in Y : h_s(y) > \alpha_s + \frac{\alpha^{n1}}{2} \},$$

is a weakly open subset of V and contains y_{n1}. Thus, $\tilde{V} \cap H_n \neq \emptyset$ and

$$diam(\tilde{V} \cap H_n) \leq diam(V \cap H_n) < \varepsilon.$$

We denote $H_n^1 = \tilde{V} \cap H_n$ and assign to it the integer

$$j(H_n^1) = \min \{ i \in \mathbb{N} : \frac{1}{i} \leq \frac{\alpha^{n1}}{2}, \quad and \quad \varepsilon - diam(V \cap H_n) \geq \frac{1}{i} \}.$$

Now, we proceed inductively. Assume that the sets H_n^β have been obtained for all $0 \leq \beta < \xi$, and consider the set

$$R_n^\xi = H_n \backslash (\bigcup_{\beta < \xi} H_n^\beta).$$

If $R_n^\xi = \emptyset$ the process stops and we set $\xi_n = \xi$. If not, R_n^ξ is a nonempty subset of H_n and so there exists a weakly open set $V^{n\xi}$ verifying

$$V^{n\xi} \cap R_n^\xi \neq \emptyset \quad and \quad diam(V^{n\xi} \cap R_n^\xi) < \varepsilon;$$

where

$$V^{n\xi} = \bigcap_{s=1}^{k_{n\xi}} \{ y \in Y : h_s^{n\xi}(y) > \alpha_s^{n\xi} \}$$

with $\alpha_s^{n\xi}$ real numbers and $h_s^{n\xi}$ norm one elements of X^*, $s=1,\ldots, k_{n\xi}$.

If some $y_{n\xi} \in V^{n\xi} \cap R_n^\xi$ is fixed then

$$\alpha^{n\xi} = \min \{ h_s^{n\xi}(y_{n\xi}) - \alpha_s^{n\xi} : s=1,\ldots, k_{n\xi} \}$$

is positive; hence the set

$$\tilde{V}^{n\xi} = \bigcap_{s=1}^{k_{n\xi}} \{ y \in Y : h_s^{n\xi}(y) > \alpha_s^{n\xi} + \frac{\alpha^{n\xi}}{2} \}$$

is weakly open and contains $y_{n\xi}$. Thus, the set $H_n^\xi = \tilde{V}^{n\xi} \cap R_n^\xi$ is nonempty and

$$diam(\tilde{V}^{n\xi} \cap R_n^\xi) \leq diam(V^{n\xi} \cap R_n^\xi) < \varepsilon.$$

We assign to this set the integer

$$j(H_n^\xi) = \min \{ i \in \mathbb{N} : \frac{1}{i} \leq \frac{\alpha^{n\xi}}{2}, \quad and \quad \varepsilon - diam(V^{n\xi} \cap R_n^\xi) \geq \frac{1}{i} \}.$$

In this way, we have obtained a partition $\{H_n^{\xi} : 0 \leq \xi < \xi_n\}$ of H_n and assigned to each of its members a positive integer $j(H_n^{\xi})$. Since we may assume without loss of generality that the sets H_n are disjoint, the family $\{H_n^{\xi} : 1 \leq \xi < \xi_n\}$ is a partition of Y. So, every $y \in H$ belongs exactly to one member of this partition, say $H_{n_y}^{\xi_y}$ and it makes sense to define

$$j(y) = j(H_{n_y}^{\xi_y}).$$

Set $H_n^{\nu} = \{y \in H_n : j(y) = \nu\}$. This forms a countable partition of Y, which allows us to define a countable cover $\{E_{n,l}^{\nu}\}_{n,\nu,l}$ of X in the following way:

$$E_{n,l}^{\nu} = \{x \in X : P(x) \in H_n^{\nu} \text{ and } qx \in F_l^{(\nu)}\}.$$

Indeed, given $x \in X$ there exist positive integers n, ν with $P(x) \in H_n^{\nu}$, and for this ν one has $Z = \bigcup_{k=1}^{\infty} F_k^{\nu}$, hence there exists some $l \in \mathbb{N}$ for which $qx \in F_l^{\nu}$.

We have to show that each nonempty subset of $E_{n,l}^{\nu}$ has a nonempty relatively open subset of diameter not greater than 2ε. So, we fix $n, \nu, l \in \mathbb{N}$ and a nonempty subset A of $E_{n,l}^{\nu}$ for the rest of the proof.

Since $q(A) \subset F_l^{\nu}$, there exists a weakly open set U in Z with $U \cap q(A) \neq \emptyset$ and $\text{diam } B(U \cap q(A)) < 1/\nu$. Denoting $C = A \cap q^{-1}(U)$, we see that $P(C)$ is a nonempty subset of

$$H_n^{\nu} \subset H_n = \bigcup_{\xi < \xi_n} H_n^{\xi}$$

and it is then possible to define

$$\xi_C = \min\{\xi \in [1, \xi_n) : H_n^{\xi} \cap P(C) \neq \emptyset\}.$$

We have that $P(C)$ is contained in

$$R_n^{\xi_C} = H_n \backslash (\bigcup_{\beta < \xi_C} H_n^{\beta})$$

and its intersection with $H_n^{\xi_C}$ is non empty, which yields

$$j(H_n^{\xi_C}) = \nu$$

since $P(C) \subset H_n^{\nu}$: i.e., $j(x) = \nu$ for every $x \in P(C)$.

Let

$$V = \bigcap_{s=1}^{k} \{y \in Y : h_s(y) > \alpha_s\}$$

and let

$$\tilde{V} = \bigcap_{s=1}^{k} \{y \in Y : h_s(y) > \alpha_s + \frac{\alpha}{2}\},$$

with α_s real numbers and h_s norm one elements of X^*. Let us assume that

$$H_n^{\xi C} = \tilde{V} \cap R_n^{\xi C}.$$

By construction $1/v \le \alpha/2$; and, moreover,

$$1/v \le \varepsilon - diam(\, V \cap R_n^{\xi C}).$$

Let us fix a point x in C with $P(x) \in \tilde{V}$ (which exists since

$$P(C) \cap H_n^{\xi C} = P(C) \cap \tilde{V} \ne \varnothing).$$

It is then possible to define

$$W = \{z \in X : qz \in U\} \cap \bigcap_{s=1}^{k} \{z \in X : h_s(z) > \alpha_s + \frac{\alpha}{2} + h_s(Bqx)\}.$$

Clearly W is a weakly open subset of X. Moreover,

$$A \cap W = C \cdot \cap \bigcap_{s=1}^{k} \{z \in X : h_s(z) > \alpha_s + \frac{\alpha}{2} + h_s(Bqx)\}.$$

Since x belongs to C one has

$$h_s(x) = h_s(P(x)) + h_s((Bqx) > \alpha_s + \alpha/2 + h_s(Bqx),$$

obtaining that $x \in A \cap W$. It remains to estimate the diameter of $A \cap W$. To do that, let $y \in A \cap W$. One has

$$\| x-y \| \le \| Px-Py \| + \| Bqx-Bqy \|.$$

Since $x, y \in A \cap W \subset C$ then $qx, qy \in q(A) \cap U$, and since $diam(B(q(A) \cap U) \le 1/v$ then $\| Bqx-Bqy \| \le 1/v$. On the other hand, $y \in W$ yields

$$\begin{aligned} h_s(Py) &= h_s(y) - h_s(Bqy) \\ &> \alpha_s + \alpha/2 + h_s(Bqx) - h_s(Bqy) \\ &= \alpha_s + \alpha/2 + h_s(Bqx - Bqy). \end{aligned}$$

But since

$$|h_s(Bqx - Bqy)| \le \| Bqx-Bqy \| \le 1/v,$$

by the choice $v = j(H_n^{\xi C})$ we obtain (for $s = 1, \dots, k$)

$$h_s(Py) > \alpha_s + \alpha/2 - 1/v \ge \alpha_s,$$

and thus $Py \in V$. Since

$$P(y) \in P(A \cap W) \subset P(C) \subset R_n^{\xi C},$$

then $P(y) \in V \cap R_n^{\xi C}$; analogously

$$P(x) \in P(C) \subset R_n^{\xi C}$$

and $P(x) \in \tilde{V} \subset V$ give $P(x) \in V \cap R_n^{\xi C}$.

Hence,

$$\| Px - Py \| \le diam \, (V \cap R_n^{\xi C}).$$

Therefore

$$\|x-y\| \le \| Px-Py \| + \| Bqx-Bqy \|$$

$$\le \, diam \, (V \cap R_n^{\xi C}) + \frac{1}{v}$$

$$= \varepsilon + \frac{1}{v} - (\varepsilon - diam \, (V \cap R_n^{\xi C}))$$

$$\le \varepsilon$$

by the choice of v. Since the point y was arbitrary, $diam(A \cap W) \le 2\varepsilon$. □

The preceding result yields that σ-fragmentable spaces do not necessarily admit a *PRI*, as the example *JL* shows.

A Banach space X is said to *admit weakly countable covers of small local diameter* if for each $\varepsilon > 0$ it is possible to write X as a countable union $\cup E_n$ of sets such that every point in E_n belongs to a relatively weakly open subset of E_n having diameter less than ε. A proof analogous to the preceding one gives.

Theorem 4.16.b. *Admitting weakly countable covers of small local diameter is a 3SP property.*

Proof. It is enough to observe that in the previous proof the subsets E_{nl}^j form a weakly countable cover of small local diameter for X. Note that in this case, the construction of the partitions $\{ H_n^\xi : 0 \le \xi < \xi_n \}$ can be avoided, obtaining a simpler proof. □

Remark. Arguing as in the previous theorem, and using the norm continuity of the Bartle-Graves selection it is possible to show that admitting countable covers of sets of small local diameter that are norm closed (or differences of norm closed sets) is a 3SP property.

4.17   Point of continuity properties

A non-empty closed bounded subset K of a Banach space X is said to have the *Point-of-Continuity-Property* (PCP) if there is some point $p \in K$ where the identity

$$(K, \ weak) \to (K, \ norm)$$

is continuous [39]. Equivalently [39, Lemma 2], for every $\varepsilon > 0$ there is a weakly open set U with

$$U \cap K \neq \emptyset \text{ and } diam(U \cap K) < \varepsilon.$$

The Banach space X satisfies the *PCP* when every closed subset K of X has *PCP*. This property has been introduced in [39].

The 3*SP* problem for the *PCP* was posed by Edgar and Wheeler [108], who proved (Corol. 5.3.) the following restricted form of the 3*SP* property: *If X has Cech complete ball and X^{**}/X has PCP, then X^{**} has PCP.* Rosenthal [315] solved the problem with a positive answer.

Theorem 4.17.a. *The PCP is a 3SP property.*

Proof. The proof follows the schema: first assume it is true for the separable case and prove that is true in general; then prove the separable case.

STEP 1. *If every closed separable subset of X has PCP then every closed subset of X has PCP.*

Proof. For if K is a closed subset of X failing *PCP* then there is some closed subset A of K and some $\delta > 0$ such that for every non empty relatively weakly open subset $W \subset A$, *diam* $W > \delta$. It follows that for every point $a \in A$ one has

$$a \in \overline{\{x \in A : \|x-a\| > \delta/2 \}}^{weak}.$$

By a result of Kaplansky (see Floret [123]),

$$\forall a \in A \ \exists \ A_a \subset \{x \in A : \|x-a\| > \delta/2 \}, \quad card \, A_a = \mathbb{N} : \quad a \in \overline{A_a}^{weak}.$$

So, if $a_0 \in A$ is fixed and one sets $A_1 = \{a_0\} \cup A_{a_0}$ and proceed inductively choosing countable subsets $A_n \subset A$ such that

$$A_{n+1} = A_n \cup \bigcup_{a \in A_n} A_a$$

then the set $L = \cup A_n$ is countable. Moreover, $\bar{L}$ fails PCP because if U is a weakly relatively open subset of L and we choose n and $a \in A_n \cap U$ such that $a \in \bar{A_a}^{\text{weak}}$ then there is some $x \in A_a \cap U \subset L \cap U$. Hence, $\|x-a\| > \delta/2$.   □

STEP 2. *If the PCP is a 3SP property restricted to the class of separable spaces then it is a 3SP property.*

Assume that some Banach space X fails the PCP while some subspace Y and the corresponding quotient space X/Y have the PCP. By the result of Step 1, some separable subspace B of X fails the PCP. For the next lemma, recall that $q: X \to X/Y$ is the quotient map.

Lemma 4.17.b. *Let Y and B be closed subspaces of a Banach space X, with B separable. There exist separable subspaces X_1 and Y_1 of X with $Y_1 \subset X_1 \cap Y$ such that $B \subset X_1$ and $\overline{qB}^{X/Y}$ is isometric to X_1/Y_1.*

Proof. The subspace Y_1 is chosen as a separable subspace of Y such that for all points $b \in B$ one has

$$dist\ (b,\ Y_1) = dist\ (b,\ Y),$$

This can be done since B is separable. The subspace X_1 is chosen as the closure of $Y_1 + B$. Denote by $q_1 : X \to X/Y_1$ the corresponding quotient map and define an operator $T: q_1(B) \to q(B)$ by $Tq_1 b = qb$. The operator T is clearly well-defined and surjective. The choice of Y_1 implies that T is an isometry and can thus be extended to a surjective isometry

$$\overline{qB}^{X/Y} \to \overline{q_1 B}^{X_1/Y_1} = X_1/Y_1 .   \qquad \square$$

So, there are separable subspaces X_1 and Y_1 of X such that $Y_1 \subset Y$, $Y_1 \subset X_1$, $B \subset X_1$ and $\overline{qB}$ is isometric to X_1/Y_1. Since Y_1 is a subspace of Y, it has the PCP; since $\overline{qB}$ is a subspace of X/Y it also has the PCP. Therefore X_1/Y_1 must also have the PCP. This, and the hypothesis yield that X_1 has the PCP as well as B. A contradiction.   □

The proof for the separable case comes next.

LAST STEP. *The PCP is a 3SP property restricted to the class of separable spaces*

Proof. Assume that Y is a subspace of a separable space X such that Y and X/Y have the PCP while X fails the PCP. Then, for some $\delta > 0$ and some subset $A \subset Ball(X)$ and for every weakly open set U such that $U \cap A \neq \emptyset$, one has $diam\ (U \cap A) > \delta$.

New ideas, essential for the remainder of the proof, emerged from Bourgain and Rosenthal [39] (if part of the next proposition) and Ghoussoub and Maurey [128] (only if part).

Some notation is needed. Given a sequence (G_i) of subspaces of a Banach space X we denote $[G_i]_{i=1}^\infty$ the closed subspace of X generated by the G_i, and for $1 \le m \le n \le \infty$ we denote $G[m,n]$ the subspace generated by $\{G_i: m \le i \le n\}$. With all this, let P be a class of finite dimensional decompositions (*FDD*). A sequence (G_i) of finite dimensional subspaces of X is a *P-skipped blocking decomposition* (*P-SBD*) of X if the following conditions hold

i) $X = [G_i]_{i=1}^\infty$

ii) $G_i \cap [G_j]_{j \ne i} = \{0\}$ for all i

iii) If (m_k) and (n_k) are sequences of integers, with $m_k < n_k < m_{k+1}$ for every k, then the sequence $\{[G\ [m_k,n_k]\ ]\}_{k=1}^\infty$ is a *FDD* of $[G\ [m_k,n_k]]_{k=1}^\infty$ of class P.

Proposition 4.17.c. *A Banach space has the PCP if and only if it admits a boundedly complete skipped blocking decomposition.*

Proof. Let (G_j) be a boundedly complete skipped blocking decomposition for Y, and let us choose a (unique) sequence (H_j) of finite dimensional subspaces of Y^* biorthogonal to (G_j); i.e.,

$$H_j = \left([G_j]_{j \ne i}\right)^\perp.$$

Let $\varepsilon_j = 2^{-j}$. We shall inductively construct two sequences (m_j) and (n_j) of integers and two sequences (x_j) in A and (y_j) in X satisfying, for all j,

(i) $m_j < n_j + 1 < m_{j+1}$

(ii) $\|x_{j+1} - x_j\| > \delta/2$

(iii) $y_j \in G[m_j, n_j]$ and $\|(x_{j+1} - x_j) - y_j\| < 3\varepsilon_j$.

This implies that $F_j = G[m_j, n_j]$ is a skipped blocking decomposition of $[F_j]_j$ that is not boundedly complete, since $y_j \in F_j$ and, while

$$\|\Sigma\ y_j\| \le 5,$$

the series Σy_j does not converge because

$$\lim \inf \|\ y_j\ \| \ge \delta/2.$$

To start the construction we need a lemma translating the *PCP* from X/Y:

Lemma 4.17.d. *If X/Y has the PCP then for every bounded subset L of X, every relatively weakly open subset W of L and every $\varepsilon > 0$ there exists a weakly open subset U of X such that $\varnothing \neq U \cap L \subset W$ and diam $q(U \cap L) < \varepsilon$.*

Proof. Given L, W and ε as indicated, we can choose a weakly open subset Λ in X/Y such that $\Lambda \cap q(W) \neq \varnothing$ and $diam(\Lambda \cap q(W)) < \varepsilon$. Choosing V a weakly open subset of X such that $V \cap L = W$ and putting $U = V \cap q^{-1}(\Lambda)$ it turns out that

$$diam\ q(V \cap q^{-1}(\Lambda)) = diam\ (q(V) \cap \Lambda) = diam\ (q(W) \cap \Lambda) < \varepsilon$$

and

$$\varnothing \neq U \cap L = V \cap L \cap q^{-1}(\Lambda) = W \cap q^{-1}(\Lambda) \subset W \qquad \square$$

Initial step. By the lemma, let U_1 be a weakly open subset of X with $U_1 \cap A \neq \varnothing$ and *diam* $q(U_1 \subset A) < \varepsilon_1$. Pick $x_1 \in U_1 \cap A$.

Choose now U_2 such that $\varnothing \neq U_2 \cap A \subset U_1 \cap A$ and *diam* $q(U_2 \cap A) < \varepsilon_2$. Pick $x_2 \in U_2 \cap A$ such that $\|x_2 - x_1\| > \delta/2$. Since *diam* $q(U_1 \cap A) < \varepsilon_1$, $\|qx_2 - qx\| < \varepsilon_1$. Recalling that $Y = [G_i]$ it is possible to pick $y_1 \in [G_i]$ such that $\|(x_2 - x_1) - y_1\| < \varepsilon_1$. We set $m_1 = 1$ and n_1 such that $y_1 \in G[m_1, n_1]$.

Inductive argument. Assume that m_{j-1}, n_{j-1}, $x_j \in A$ and a weakly open neighborhood U_j of x_j have been chosen so that *diam* $q(U_j \cap A) < \varepsilon_j$.

We set $m_j = n_{j-1} + 2$ and $F_j = H[1, m_j - 1] = G[m_j, \infty]$. The following result shows the way to continue the induction.

Lemma 4.17.e. *Let Y be a subspace of X and let F be a finite dimensional subspace of Y^*. Given $\varepsilon > 0$ there exists a weakly open subset W of X such that for every $w \in W$, dist $(w, Y) < \varepsilon \Rightarrow$ dist $(w, {}^{\perp}F) < 3\varepsilon$.*

Proof. Let X, Y, F and ε be as in the statement of the lemma. Choose $\tau > 1$ and $\eta > 0$ so that $\tau(\varepsilon + \tau) \leq 2\varepsilon$.

Claim. There exists a finite dimensional subspace G of X^* such that for all $h \in X^*$ with $h|_Y \in F$ there exist $g \in G$ and $y^{\perp} \in Y^{\perp}$ with $\|g\| \leq \tau\|h\|$ and $h = g + y^{\perp}$.

Proof of the claim: Since *Ball*(F) is compact, it is possible to choose a finite number of points $w_1, \dots, w_n$ in $\tau Ball(F)$ in such a way that $Ball(F) \subset conv\{w_1, \dots, w_n\}$. Let W_i be a norm-preserving Hahn-Banach extension of w_i to X^*. The space G is the span of $\{W_1, \dots, W_n\}$. Because if h is as in the statement of the claim, the restriction of $h/\|h\|$ to Y is a convex combination of w_i; if $g/\|h\|$ is the same convex combination of W_i then $\|g/\|h\|\| \leq \|w_i\| \leq \tau$. Moreover g and h differ in an element of $Y^{\perp}$. $\qquad \square$

So, let G a finite dimensional space as in the claim, and define

$$W=\{x \in X : \| x|_G \| <\eta\}.$$

It is easy to verify that W is a weakly open neighborhood of zero: The set $X\backslash W$ is weakly closed since every net (z_α) of elements of $X^{**}\backslash W$ weak* convergent to z satisfies that $z_\alpha|_G$ is also weak convergent (and thus norm convergent) to $z|_G$. Since $\| z_\alpha|_G \| \geq\eta$, the same happens to z.

Let then $w \in W$ with $dist(w, Y)<\varepsilon$, and let $d=dist(w, {}^\perp F)$. By the Hahn-Banach theorem it is possible to choose a norm one element $h \in X^*$ with $h(w)=d$ and $h({}^\perp F)=0$. Hence, $h|_Y \in ({}^\perp F)^\perp \cap Y = F$.

After the claim, there is a $g \in G$ with $\| g \| \leq\tau$ and $g-h = y^\perp \in Y^\perp$. Since $w \in W$ and $\| g \| \leq\tau$, one has $|g(w)|\leq\tau\eta$. Hence,

$$d = h(w) = g(w) + y^\perp(w)$$

$$\leq |g(w)| + |y^\perp(w)|$$

$$< \tau\eta + \|y^\perp\| \, dist(w,Y)$$

$$< \tau\eta + (1+\tau)\varepsilon ,$$

and it follows that $d<3\varepsilon$. $\qquad\square$

End of the proof of the Last Step of 4.17.a. Choose W_j a weakly open neighborhood of 0 in X such that for all $w \in W_j$

$$dist (w,Y) <\varepsilon_j \Rightarrow dist (w,F_j^\perp) <3\varepsilon_j$$

By Lemma 4.17.e, it is possible to choose a weakly open subset U_{j+1} of X such that

$$\emptyset \neq U_{j+1} \cap A \subset U_j \cap A \cap (x_j + W_j)$$

and

$$diam \, q(U_{j+1} \cap A)<\varepsilon_{j+1}.$$

It is possible to pick $x_{j+1} \in U_{j+1} \cap A$ such that $\|x_{j+1}-x_j\| >\delta/2$. Since $x_{j+1} \in U_j \cap A$, and applying the induction hypothesis, one obtains

$$\| q(x_{j+1})-q(x_j)\| <\varepsilon_j;$$

thus

$$dist (x_{j+1}-x_j, Y)<\varepsilon_j.$$

Moreover (by choice) $x_{j+1}-x_j \in W_j$ and this yields

$$dist\ (x_{j+1}-x_j, F_j^\perp) < 3\varepsilon_j.$$

So, we can choose $y_j \in G[m_j, \infty]$ such that $\|x_{j+1}-x_j-y_j\| < 3\varepsilon_j$. The process finishes choosing n_j such that $y_j \in G[m_j, n_j]$.   □

A Banach space is said to have the *Convex point of continuity property* if convex closed bounded sets have *PCP*. A dual space is said to have *C* point of continuity property* if convex closed bounded sets have *PCP* with respect to the weak* topology.

Problems. *Are the C*PCP and the CPCP 3SP properties?*

(these problems appear explicitly posed in [173]).

In [18] it is considered the 3*SP* character of the so-called *property II* (introduced in [69]): the set of points of weak*-to-norm continuity of the unit sphere of X^* is norm dense in the unit sphere of X^*. This property is a weakening of the *Namioka-Phelps property*: weak* and norm topologies coincide in the unit sphere of X^*. They prove:

Theorem 4.17.f. *Property II is not a 3SP property.*

Proof. What Basu and Rao prove in [18] is that property II is not stable by products in very much the way that was done with the Hahn-Banach smooth property in 1.9.f. The key is now to prove that (y^*, z^*) is a w^*-*PC* of the unit sphere of $(Y\oplus_1 Z)^*$ if and only if $\|y^*\| = 1 = \|z^*\|$ and y^* and z^* are w^*-*PC* points of the unit spheres of Y^* and Z^* respectively. Hence, $(Y\oplus_1 X)^*$ never has property II.□

Observe that the proof also yields that infinite dimensional spaces with property II do not admit nontrivial *L*-projections.

4.18   Property (C)

Corson introduced in [74] the following property, denoted property (*C*) after [296]: A convex subset M of a Banach space is said to have *property* (*C*) if every collection of closed convex subsets of M with empty intersection contains a countable subcollection with empty intersection.

Lemma 4.18.a. *If X fails property (C), then there exist a collection $\mathscr{E}$ of nonempty convex subsets of the open unit ball B_X of X that is closed under countable intersections, and $\varepsilon > 0$ so that for every convex subset M of X with* *property (C) there is* $C_M \in \mathcal{E}$ *with dist(M, C_M)$\geq \varepsilon$.*

Proof. Assume that X fails property (*C*), and take a collection $\mathcal{E}$ of closed convex subsets of E with empty intersection, but such that every countable subcollection has nonempty intersection. Note that adding to $\mathcal{E}$ the intersections of countable subcollections, one obtains a collection that is closed under countable intersections. Thus, there is no loss of generality assuming that $\mathcal{E}$ is closed under countable intersections.

Claim. There exists a collection $\mathfrak{Z}$ of nonempty convex sets in X closed under countable intersections and $\sigma > 0$ so that for every $a \in X$ there is some $C_a \in \mathfrak{Z}$ with dist(a, C_a)$>\sigma$.

Proof of the claim. Otherwise, select inductively for $i=1, 2,...$ and $C^\circ \in \mathcal{E}$ a nonempty convex set C^i and some $a_i \in X$ in such a way that

    i) $C^\circ \supset C^1 \supset C^2 \supset ...$; with *diam* $C^{i+1} \leq 2^{-i}$
    ii) The class $\mathcal{E}^i = \{C^i : C^\circ \in \mathcal{E}\}$ is closed under countable intersections
    iii) dist $(a_i, C^i) < 2^{-i-1}$.

This can be done as follows: first, choose $a_0 \in E$ such that for all $C \in \mathcal{E}$ one has: *dist* $(a_0, C) \leq 2^{-2}$; then, assuming that a_i and $\mathcal{E}^i$ have been defined, put

$$C^{i+1} = C^i \cap (a_i + 2^{-i-1}B_E).$$

The set C^{i+1} is nonempty by *iii)*. Condition *i)* is obvious, and *ii)* follows from the same condition for C^i. Finally, select a_{i+1} verifying *iii)* using the negation of the claim for C^{i+1} and $\sigma = 2^{-i-3}$.

Now, one has $\| a_i - a_{i+1} \| \leq 2^{-i}3$; and so *lim* $a_i = a \in E$. By *iii)*, *dist(a, C)$=0$* for all $C \in \mathcal{E}$. Since C is closed, $a \in \cap_{C \in \mathcal{E}} C$. A contradiction that proves the claim.   □

Take then the collection $\mathfrak{Z}$ and $\sigma > 0$ given in the claim. Since $\mathfrak{Z}$ is closed under countable intersections, there is $n \in N$ such that $C \cap nB_E \neq \emptyset$ for every $C \in \mathfrak{Z}$.

Put

$$\mathfrak{C} = \left\{ \frac{1}{n} (C \cap nB_E) : C \in \mathfrak{Z} \right\}$$

and $\varepsilon = \sigma/n$

Clearly $\mathfrak{C}$ consists of nonempty convex subsets of B_E, it is closed under countable intersections, and for every $x \in E$ there exists $C_x \in \mathfrak{C}$ such that

$$dist (x, C_x) > \varepsilon.$$

Moreover, if M is a convex set with property (C), and $dist(M, C) < \varepsilon$ for every $C \in \mathfrak{C}$, taking $C^* = (C + \varepsilon B) \cap M$ and letting C run through $\mathfrak{C}$ one obtains a family $\{C^* : C \in \mathfrak{C}\}$ of nonempty convex subsets of M that is closed under countable intersections. By property (C) of M there exists

$$x \in \cap_{C \in \mathfrak{C}} C^*,$$

which yields $dist(x, C) < \varepsilon$ for every $C \in \mathfrak{C}$. A contradiction that completes the proof. $\square$

Theorem 4.18.b.[296]. *Property (C) is a 3SP property.*

Proof. Let $0 \to Y \to X \to Z \to 0$ be a short exact sequence with Y and Z having property (C). Take a collection $\mathfrak{C}$ of subsets of X as in the lemma. For every $x \in X$, $x + Y$ is a translate of Y and thus has property (C). Then, there is $C_x \in \mathfrak{C}$ with

$$dist\,(x + Y, C_x) \geq \varepsilon.$$

This implies

$$dist\,(qx, qC_x) \geq \varepsilon.$$

Thus $qx \notin \overline{qC_x}$ and hence

$$\cap_{x \in X} \overline{qC_x} = \varnothing.$$

However, for every countable set $\{x_n\} \subset X$ one has

$$C = \cap_{n \in N} C_{x_n} \in \mathfrak{C};$$

so $C \neq \varnothing$ and

$$\varnothing \neq qC \subset \cap_{n \in N} \overline{qC_{x_n}},$$

in contradiction with property (C) of X/Y. $\square$

Theorem 4.18.c. *Property $(C)^{dual}$ is a 3SP property.*

In [132], Godefroy proved that, for separable spaces, $(C)^{dual}$ is equivalent to not containing l_1.

4.19 GSG spaces

A Banach space X is said to be a *GSG* space if there is an Asplund space A and an operator $T{:}A \to X$ with dense range. These spaces were studied in [331]. Asplund spaces and *WCG* spaces are *GSG*.

Theorem 4.19.a. *GSG is not a 3SP property.*

Proof. The sequence $0 \to C[0, 1] \to D[0, 1] \to c_0(I) \to 0$ gives a counterexample, since [331, Thm. 3.1] implies that *GSG* spaces are weakly Asplund. See now 5.6.   □

Some results in [80, p. 192] are relevant to the study of 3SP problems related to *GSG* spaces, so we describe them.

Theorem 4.19.b. *The property "the unit dual ball does not contain a homeomorphic copy of* $[0, \omega_1]$*" is a 3SP property.*

Proof. Let $0 \to Y \to X \to Z \to 0$ be a short exact sequence and assume that *Ball*(Y^*) and *Ball*(Z^*) do not contain $[0, \omega_1]$. Let $i^*: X^* \to Y^*$ be the natural quotient map, and assuming $[0, \omega_1]$ is homeomorphic to a subset K of *Ball*(X^*), since *Ker* i^* does not contain $[0, \omega_1]$, for every $y \in i^*(K)$ one has that $(i^*)^{-1}(y) \cap K$ is countable. It is easy to construct by transfinite induction a cofinal closed subset F of K such that $i^*|_F$ is one-to-one; hence $i^*(F)$ is homeomorphic to $[0, \omega_1]$; a contradiction.   □

This property is the key to obtain (see [80])

Proposition 4.19.c. *Let X be an Asplund space such that* X^* *is GSG. Then either Ball*(X^{**}) *contains* $[0, \omega_1]$ *or* X^* *is WCG*

and thus arrive to

Theorem 4.19.d. *{Asplund and WCG}^{dual} is a 3SP property.*

Proof. Assume Y^* and $(X/Y)^*$ are Asplund and *WCG*. Since Y^* and $(X/Y)^*$ are *WCG* then Y and X/Y are Asplund (4.11.b); hence X is Asplund. Since Y^* and $(X/Y)^*$ are Asplund spaces, X^* is an Asplund space and thus *GSG*. Since Y^* and $(X/Y)^*$ are *WCG* their unit balls do not contain $[0, \omega_1]$ and, by 4.19.b, the same happens to the unit ball of X^*. Applying 4.19.c one obtains that X^* is *WCG*.   □

However, the sequence $0 \to c_0 \to JL \to l_2(\Gamma) \to 0$ shows

Theorem 4.19.e. *{Asplund and WCG} is not a 3SP property.*

Question. *Is the property* " *(Ball*(X^*), w^*) *does not contain an homeomorphic copy of* $\beta\mathbb{N}$ " *a 3SP property?* By a result of Talagrand (see [340]) this property is equivalent to "not having l_∞ as a quotient" (see also the second question at 3.2.g).

Appendix 4.20

Polynomial properties

The study of polynomials in Banach space still has not many three-space results available. We can, however, formulate some of the main properties involving polynomials and look for the solution of their $3SP$ problems

Weakly sequentially continuous polynomials

Continuous polynomials tend to be not weakly continuous as the square of the norm of a Hilbert space clearly shows. In fact, the property that all polynomials are weakly sequentially continuous seems to be quite "pathological." Let us call it property *pwsc*. Known spaces with property *pwsc* are those with the Dunford-Pettis property [316]. The *hereditary pwsc* property means that all subspaces have *pwsc*. The known examples of spaces having *pwsc* hereditarily are thus the spaces with the hereditary Dunford-Pettis property and the spaces where no normalized sequence admitting a lower p-estimate exists [151].

Theorem 4.20.a. *The properties pwsc and hereditary pwsc are not $3SP$ properties.*

Proof. Consider the sequence $0 \to Ker\,(q+i) \to l_1 \oplus S \to c_0 \to 0$ as described in 4.6.a and 6.6.f. One only needs to prove that Schreier's space S has not property *pwsc*. If (e_n) denotes the canonical basis, the estimate

$$\left\| \sum_{j=1}^{N} e_{n_j}^* \right\|_{S^*} \leq 2\,logN,$$

obtained in 6.6.f for the canonical basis of S^* and some standard duality results yield that the canonical basis of S, which is a normalized weakly null sequence, satisfies moreover a lower p-estimate for all $p > 1$. Thus the map $p(\Sigma\lambda_n e_n) = \Sigma(\lambda_n)^2$ is a continuous 2-homogeneous polynomial not weakly sequentially continuous since $p(e_n) = 1$. $\square$

R. Gonzalo and R. García mentioned that the following partial improvement is possible: there exist Banach spaces X, Y where all 2-polynomials are weakly sequentially continuous while there exist 2-polynomials in $X \oplus Y$ that are not weakly sequentially continuous.

Spaces admitting a separating polynomial

In this section we consider real Banach spaces. A polynomial P on a Banach space X is said to be *separating* if it satisfies $P(0)=0$ and $|P(x)|>1$ for all x in the unit sphere of X. Hilbert spaces, and in general $L_{2n}(\mu)$-spaces ($n \in \mathbb{N}$), admit a separating polynomial; namely, $P(f) = \|f\|^{2n}, f \in L_{2n}(\mu)$

Theorem 4.20.b. *To admit a separating polynomial is not a 3SP property.*

Proof. It is enough to show that the space Z_2 does not admit a separating polynomial. This follows from two facts:

i) The only Banach spaces with symmetric basis that admit a separating polynomial are the spaces l_{2n}, $n \in N$. [142]

ii) The space Z_2 contains a copy of the Orlicz space l_M, with $m(x)=x^2(\log(x))^2$.

□

Problem. In several papers [49, 143, 159, 180], polynomial versions of properties defined with operator ideals have been introduced and studied: polynomial (V), polynomial C_1, polynomial Schur (also called Λ-spaces)... A Banach space X is said to be a Λ-space [49] if polynomially null sequences are norm null. Jaramillo and Prieto [180] proved that spaces in $(W_p)^{dual}$ are Λ-spaces, setting in the affirmative a conjecture of Carne, Cole and Gamelin [49]. It is apparently an open question whether *to be a Λ-space* is a 3SP property. The difficulty to make a direct proof via lifting analogous to that of Schur property is that polynomials cannot be extended from subspaces to the whole space. Moreover, new polynomial properties such as Q-reflexivity (see [10, 141]) and other ([181, 351] are being investigated by many authors. An excellent updated survey is [159].

It would be interesting to obtain 3SP results in this context.

Chapter 5

Geometrical Properties

Many properties of Banach spaces could be described as "geometrical." In something like an exercise in schizophrenia, during a part of this chapter we focus on those that come defined in terms of differentiability properties of the given norm (and/or its dual norm) ... to study the 3*SP* problem for *having an equivalent norm with that property*. A typical example is the geometrical property of uniform convexity whose topological equivalent is super-reflexivity. A lot of research has been done trying to describe the Banach spaces that admit an equivalent norm (a *renorming*, in short) with some good geometrical properties. Classical results can be found in chapter 4 of Diestel's book [89]; and an up-to-date comprehensive reference is the masterpiece [82] of Deville, Godefroy and Zizler, to which we shall often refer to. The chapter includes (a sketch of) the (quite a few) positive results available, such as the 3*SP* property for admitting a locally uniformly convex renorming of Godefroy, Troyanski, Whitfield and Zizler [136]; nonetheless, the highlights of this topic are the examples: Talagrand's counterexample to the 3*SP* property [341] for admitting a (uniformly) Gâteaux smooth renorming, or Haydon's counterexample [164] for admitting a strictly convex renorming. One of the main intriguing open questions in this area is the 3*SP* problem for Fréchet smooth renorming.

The study of the (still open) problem of the equivalence between the Radon-Nikodym and Krein-Milman properties has produced an intensive development of the geometric theory of Banach spaces. Several properties and notions, such as extreme point, exposed point, dentability or slices, emerge from that study; and have been related to differentiability of measures, convergence of martingales and other analytical concepts. Other properties that can adopt these forms, such as weak Asplund or U.M.D., are also considered in the chapter. However, organization surprises continue since the Radon-Nikodym property shall not be studied here; we postpone it until Chapter 6 because its characterization in terms of operators is more suitable to the study of the 3*SP* problems.

5.1   Type and cotype

We refer to Chapters 11 and 12 in [90] where all one would like to know about type and cotype can be found. A Banach space X is said to have *type* p, $1 \leq p \leq 2$, if there is a positive constant C such that for every finite set (x_i) one has

$$\text{Average} \left\| \sum^N \pm x_i \right\| \leq C \left(\sum^N \|x_i\|^p \right)^{1/p}.$$

All spaces have type 1, Hilbert spaces have type 2 and no space can have type $p > 2$. A Banach space X is said to have *cotype* q, $2 \leq q \leq \infty$, if there is a positive constant C such that for every finite set (x_i) one has

$$C \left(\sum^N \|x_i\|^q \right)^{1/q} \leq \text{Average} \left\| \sum^N \pm x_i \right\|.$$

All spaces have cotype ∞, Hilbert spaces have cotype 2 and no space can have cotype $q < 2$. In general, there is no duality between type and cotype: l_1 has cotype 2 while l_∞ has no nontrivial type or cotype. For B-convex spaces one has (see [90, 13.17]):

Proposition 5.1.a. *Assume that X is a B-convex space. X has cotype q if and only if X^* has type q^*.*

Regarding $3SP$ questions, one has.

Theorem 5.1.b. *To have type p or cotype q are not $3SP$ properties.*

Proof. The sequence $0 \to l_p \to Z_p \to l_p \to 0$ for $p \leq 2$ gives a counterexample to the type p assertion. Recall from 4.6.e that

$$F \left(\sum_{i=1}^N \pm e_i \right) = (\log \sqrt[p]{N}) \sum_{i=1}^N \pm e_i$$

which gives

$$\left\| (0, \sum^N \pm e_j) \right\| \geq \sqrt[p]{N} \, (\log \sqrt[p]{N})$$

It is therefore impossible the existence of a constant $c >$ such that

$$\text{Average} \left\| \sum^N (0, \pm e_i) \right\| \leq c \left(\sum^N \|(0, e_i)\|^p \right)^{1/p} = c \sqrt[p]{N}.$$

and thus Z_p has not type p.

The proof for cotype follows by duality and 5.1.a: B-convexity is a $3SP$ property and l_p spaces are B-convex for $1 < p < \infty$. Therefore the exact sequence $0 \to l_{p*} \to (Z_p)^* \to l_{p*} \to 0$ shows that $(Z_p)^*$ is B-convex and cannot have cotype p^* since Z_p has not type p. $\qquad\square$

Nevertheless, Kalton [204] obtained the following positive result.

Proposition 5.1.c. *Let* $0 \to Y \to X \to Z \to 0$ *be a short exact sequence. If Y and Z have type p then X has type r for all $r < p$.*

Proof. Recall the estimate

$$\Delta(x_1, \ldots, x_n) = \left\| F\left(\sum_{i=1}^n x_i\right) - \sum_{i=1}^n F(x_i) \right\| \leq K\left(\sum_{i=1}^n i \, \|x_i\|\right).$$

The key of the argumentation is to prove that in a type p space one has

$$\left(\int_0^1 \Delta\left(r_1(t)x_1, \ldots, r_n(t)x_n\right)^q dt\right)^{1/q} \leq D\left(\sum_{i=1}^n \|x_i\|^q\right)^{1/q}$$

for all $q < p$ ([204, pp. 264-265]), where (r_i) is the sequence of Rademacher functions. With this estimate, the remainder is easy:

$$\left(\int_0^1 \sum_{i=1}^n \|r_i(t)(y_i, z_i)\|^q dt\right)^{1/q} =$$

$$= \left[\int_0^1 \left\| \left(\sum_{i=1}^n r_i(t)y_i, \sum_{i=1}^n r_i(t)z_i\right) \right\|^q dt\right]^{1/q}$$

$$= \left[\int_0^1 \left[\left\|\sum_{i=1}^n r_i(t)y_i - F\left(\sum_{i=1}^n r_i(t)z_i\right)\right\| + \left\|\sum_{i=1}^n r_i(t)z_i\right\| \right]^q dt\right]^{1/q}$$

$$\leq \left[\int_0^1 \left(\left\|\sum_{i=1}^n r_i(t)(y_i - F(z_i))\right\| + \Delta(r_1(t)z_1, \ldots, r_n(t)z_n) + \left\|\sum_{i=1}^n r_i(t)z_i\right\|\right)^q dt\right]^{1/q}$$

$$\leq T_Y\left(\sum_{i=1}^n \|y_i - F(y_i)\|^q\right)^{1/q} + (D + T_Z)\left(\sum_{i=1}^n \|z_i\|^q\right)^{1/q}$$

$$\leq (T_Y + D + T_Z)\left(\sum_{i=1}^n (\|y_i - F(y_i)\| + \|z_i\|)^q\right)^{1/q}$$

$$= (T_Y + D + T_Z)\left(\sum_{i=1}^n \|(y_i, z_i)\|^q\right)^{1/q}.$$

$\qquad\square$

5.2 Basic results on renormings and differentiability of the norm

In this section we include some general results on extension and lifting of norms and the different properties of differentiability of the norm.

Extension of norms 5.2.a. *Let Y be a subspace of a Banach space X. Any equivalent norm* | | *on Y can be extended to an equivalent norm on X.*

Proof ([82, II.8.1]). If $\| \quad \|$ is the original norm of X, we assume that $| \quad | \leq \| \quad \|$. Thus, the Minkowski functional of

$$B = conv \left\{ Ball_{\| \quad \|} (X) \cup Ball_{| \quad |} (Y) \right\}$$

gives an equivalent norm on X that extends $| \quad |$. $\qquad\qquad\square$

Lifting of norms 5.2.b. *Let Y be a subspace of a Banach space X. Any equivalent norm on X/Y may be obtained as the quotient norm of some equivalent norm on X.*

Proof. Let $| \quad |$ be the equivalent norm on X/Y. Assume that the quotient norm associated to the original norm $\| \quad \|$ on X is smaller than or equal to $| \quad |$ on X/Y. If q denotes the quotient map $X \to X/Y$ then the Minkowski functional of

$$q^{-1}(Ball_{| \quad |} (X/Y)) \cap Ball_{\| \quad \|} (X)$$

gives the desired norm on X. $\qquad\qquad\square$

We consider some differentiability properties of the norm and their corresponding dual notions defined in terms of convexity.

Fréchet smooth norm. Let $\| . \|$ be the norm of a Banach space X. It is said to be *Fréchet smooth* if it is Fréchet differentiable at every point $x \in X\backslash\{0\}$; i.e., if for any $x, h \in X\backslash\{0\}$, the limit

$$\lim_{t\to 0} t^{-1}(\| x+th \| - \| x \|)$$

exists and is uniform in $h \in S_X$. By homogeneity, it is enough to check for $x \in S_X$.

Gâteaux smooth norm. The norm $\| . \|$ is said to be *Gâteaux smooth* if it is Gâteaux differentiable at every point $x \in X\backslash\{0\}$; i.e., if for every $x, h \in X\backslash\{0\}$, the limit

$$\lim_{t\to 0} t^{-1}(\| x+th \| - \| x \|)$$

exists. The proof of the following result can be seen in [82, I.15].

Smullyan test on differentiability of norms 5.2.c. *Let X be a Banach space.*

i) The norm is Gâteaux differentiable at a point x, $\| x \| = 1$, *if and only if there is a unique* $f \in X^*$, $\| f \| = 1$, *such that* $f(x) = 1$.

ii) The norm is Fréchet differentiable at a point x, $\| x \| = 1$, *if and only if there is a unique* $f \in X^*$, $\| f \| = 1$, *satisfying: for every* $\varepsilon > 0$ *there exists* $\delta > 0$ *such that if* $g \in Ball(X^*)$ *and* $g(x) > 1 - \delta$ *then* $\| g - f \| < \varepsilon$.

Uniformly Fréchet smooth norm. The norm $\| \ \|$ is said to be *uniformly Fréchet smooth* if the limit

$$\lim_{t \to 0} t^{-1}(\| x + th \| - \| x \|)$$

exists for each x, $y \in S_X$ and is uniform in $(x, h) \in S_X \times S_X$.

Uniformly Gâteaux smooth norm. The norm $\| \ \|$ is said to be uniformly Gâteaux smooth if for every $h \in S_X$ the limit

$$\lim_{t \to 0} t^{-1}(\| x + th \| - \| x \|)$$

exists and is uniform in $x \in S_X$.

Proposition 5.2.d. *The norm of a Banach space is uniformly Fréchet smooth (resp. uniformly convex) if and only if its dual norm is uniformly convex (resp. uniformly smooth).*

Recall that a *bump function* on X is a real function on X with a bounded nonempty support. The smoothness of the norm implies the existence of smooth bumps:

Proposition 5.2.e. *If the norm* $\| . \|$ *of X is differentiable, then X admits a bump which is differentiable in the same sense.*

Proof. Take a C^∞-function $\tau: \mathbb{R} \to \mathbb{R}$ such that $\tau = 0$ on $(-\infty, 1] \cup [3, \infty)$ and $\tau(2) \neq 0$. Thus the function $\varphi: X \to \mathbb{R}$ defined by

$$\varphi(x) = \tau(\| x \|) \text{ for } x \in X$$

is a continuously differentiable bump function on X.   □

The converse implication is not true without restrictions. Haydon [165] exhibits a $C(K)$ space admitting a C^∞-bump and no C^∞-renorming.

5.3 Fréchet and C^k-smooth renorming

Let U be an open set in a Banach space X. A function $f: U \to \mathbb{R}$ is said to be of *class* C^k *on* U, $1 \le k \le \infty$, if it admits continuous Fréchet derivatives up to order k on U. The norm is said to be C^k-smooth if it is a C^k-smooth function on $X \backslash \{0\}$.

Problem. [82, Problem VII.5] *It is not known if admitting an equivalent Fréchet smooth norm (or a Fréchet differentiable bump) is a 3SP property.*

Nonetheless, one has

Theorem 5.3.a. *Admitting a C^k-smooth renorming, $k \ge 2$, is not a 3SP property.*

Proof. Note that any continuous twice differentiable function has its first derivative locally lipschitzian. Moreover, if both X and X^* admit bump functions with locally lipschitzian derivative then X is isomorphic to a Hilbert space [82, V.3.6]. Hence, the sequence $0 \to l_2 \to Z_2 \to l_2 \to 0$ gives a counterexample since Z_2 does not admit a C^2-bump function (recall that Z_2 and $(Z_2)^*$ are isomorphic [213]). $\square$

There exists the following partial result [136]:

Theorem 5.3.b. *Let X be a Banach space and Y a subspace of X isomorphic to $c_0(\Gamma)$. If X/Y admits an equivalent C^k-smooth norm then X admits an equivalent C^k-smooth norm.*

Proof. By a suitable renorming of X (cf. lemma 5.2.c) we can assume that Y is isometric to $c_0(\Gamma)$. The exact sequence $0 \to c_0(\Gamma) \to X \to X/Y \to 0$ induces the exact sequence $0 \to (X/Y)^* \to X^* \to l_1(\Gamma) \to 0$. Let $T: l_1(\Gamma) \to X^*$ be a linear selection for the quotient map with $\| T \| \le 1$ and let us write $f_j = Te_j$, $j \in \Gamma$.

Let $\varphi: \mathbb{R} \to \mathbb{R}$ be a C^∞-function such that

  i) $\varphi(t) = \varphi(-t)$
  ii) $0 \le \varphi(t) \le 1$; moreover, $\varphi(t) = 1$ for $|t| \le 9/8$ and $\varphi(t) = 0$ for $|t| \ge 2$;
  iii) $\varphi'(t) < 0$ for $9/8 < t < 2$;
  iv) φ is concave on the set $\{t \in \mathbb{R} : \varphi(t) \ge 1/2\}$.

Let $\alpha: \mathbb{R} \to \mathbb{R}$ be another C^∞-function such that

  i) $\alpha(t) = \alpha(-t)$
  ii) $0 \le \alpha(t) \le 1$; moreover, $\alpha(t) = 1$ for $|t| \ge 1/2$ and $\alpha(t) = 0$ for $|t| \le 1/4$;
  iii) $\alpha'(t) > 0$ for $1/4 < t < 1/2$;
  iv) α is convex on the set $\{t \in \mathbb{R} : \alpha(t) \le 1/2\}$.

Let N the original norm on X and let N_q the corresponding quotient norm on X/Y. If $\|\ \|$ is an equivalent C^k-smooth norm on X/Y, that we assume to satisfy $\|\ \| \geq N_q$, let $\psi : X/Y \rightarrow \mathbb{R}$ be the function

$$\psi(qx) = 1 - \alpha(\|\ qx\ \|).$$

Clearly, ψ is a C^k-smooth function such that

$i)\ <\psi'(z), z> \ = \ -\alpha'\,(\|\ z\ \|)\,\|\ z\ \|^2 < 0\ \ \text{for } 1/4 < \|\ z\ \| < 1/2$

$ii)\ \psi$ is concave on the set $\{z \in X/Y : \psi(z) \leq 1/2\,\}$.

Let $\Phi : X \rightarrow \mathbb{R}$ be given by:

$$\Phi(x) = \psi(qx)\ \Pi_{j\, \in\, \Gamma}\ \varphi(f_j\,(x)).$$

Claim. Φ is a well-defined C^k-function on X.

Proof of the claim. First note that if $x \in X$ and $\|\ qx\ \| < 1$ then the set

$$F = \{\,j \in \Gamma\colon |f_j\,(x)| \geq 1\,\}$$

is finite: otherwise, taking f a weak* accumulation point of $\{f_i : i \in \Gamma\}$ one has that $|f(x)| \geq 1$, which is in contradiction with $\|\ qx\ \| < 1$ because $\|\ f\ \| \leq 1$ and $f \in (X/Y)^*$ as can easily be deduced from

$$i^*f = \text{weak*-lim } i^*f_j = \text{weak*-lim } i^*Te_j = \text{weak*-lim } e_j = 0$$

and the exactness of the sequence $0 \rightarrow (X/Y)^* \rightarrow X^* \rightarrow l_1(\Gamma) \rightarrow 0$. The same argument shows that for every $x \in X$ with $\|\ qx\ \| < 1$ there is some open ball B_x centered at x and some finite subset $F[x]$ of Γ such that for $y \in B_x$

$$\Phi(y) = \psi(qy)\ \Pi_{j\, \in\, F[x]}\ \varphi(f_j\,(y));$$

since $\Phi(y) = 0$ for $\|\ qy\ \| \geq 1/2$, the claim is proved.

The next step is to show that the set

$$W = \{x \in X : \Phi(x) \geq 1/2\,\}$$

is a bounded, convex and symmetric neighborhood of zero in X:

W is a neighborhood of 0.

Indeed, the ball $(1/4)Ball_N(X)$ is contained in W because when $N(x) < 1/4$ then $\|\ qx\ \| \leq N_q(qx) \leq N(x) < 1/4$ and thus

$$\psi(qx) = 1 - \alpha(\|\ qx\ \|) = 1 = \varphi(f_j\,(x))$$

for all $j \in \Gamma$.

W is bounded.

Observe that $X^* = Y^\perp \oplus l_1(\Gamma)$, Thus, every $g \in X^*$ with $N^*(g) = 1$ can be written as

$$g = h + \Sigma_j \, a_j \, f_j$$

with $h \in (X/Y)^*$, $\| h \|^* \leq 2$ and $\Sigma |a_j| \leq 1$. On the other hand, if $x \in W$ then $\| qx \| < 1/2$ and $|f_j \, (x)| \leq 2$ for all $j \in \Gamma$. Therefore $|g(x)| \leq 3$ and W is bounded.

W is symmetric.

The functions ψ and φ are even.

W is convex.

Let $x, y \in W$ and $0 \leq t \leq 1$. It is an elementary observation that given two families (r_i) and (s_i) of positive numbers such that $\Pi r_i \geq 1/2$ and $\Pi s_i \geq 1/2$ then $\Pi(tr_i + (1-t)s_i) \geq 1/2$. From $\Phi(x) \geq 1/2$ and $\Phi(y) \geq 1/2$ one obtains that the values

$$\psi(qx), \ \psi(qy), \ \varphi(f_j(x)) \text{ and } \varphi(f_j(y))$$

are all greater than or equal to 1/2. By the concavity of ψ and φ on the sets where they take values greater than or equal to 1/2 it follows:

$$\psi(\, tqx + (1-t)qy \,) \ \geq \ t\psi(qx) + (1-t)\psi(qy)$$

$$\varphi(f_j(x + (1-t)y)) \ \geq \ t\varphi(f_j \, (x)) + (1-t)\varphi(f_j(y))$$

Hence

$$\Phi(\, tx + (1-t)y \,) \ \geq \ t\Phi(x) + (1-t)\Phi(y) \ \geq \ 1/2.$$

It has therefore been proved that the Minkowski functional $\| \ \|_W$ of W is an equivalent norm on X.

Claim. $\| \ \|_W$ *is C^k- smooth on X.*

Proof of the claim. If $x \in X$ is not zero then

$$\| x \|_W = t^{-1} \text{ if and only if } \Phi(tx) = 1/2.$$

Consider the open sets $U = X \backslash \{0\}$ and $V = (0, \infty)$ and the function $F: U \times V \to \mathbb{R}$ given by $F(x, t) = \Phi(t^{-1}x) - 1/2$. Clearly F is a function of class C^k on $U \times V$. The partial derivative $\partial_t F: U \times V \to \mathbb{R}$ with respect to t is given by

$$\partial_t F(x, t) = \langle \Phi'(t^{-1}x), t^{-2}x \rangle.$$

Subclaim. For every $x \in U$ *one has*

$$\partial_t F(x, \|x\|_W) \neq 0$$

Proof of the subclaim. Let us denote $\lambda = (\|x\|_W)^{-1}$. One has

$$\Phi(\lambda x) = \psi(\lambda q x) \, \pi_{j \in \Gamma} \, \varphi(\lambda f_j(x)) = 1/2.$$

Thus, either $1/4 < \|\lambda q x\| < 1/2$ or $1 < |\lambda f_j(x)| < 2$ for at least one $j \in \Gamma$. Since ψ and φ have negative derivatives in those intervals, from the definition of ψ and $\langle \|qx\|', qx \rangle = \|qx\|^2$ it follows

$$\langle \Phi'(\lambda x), \lambda^2 x \rangle = -\alpha'(\|\lambda q x\|) \lambda^3 \|qx\|^2 \prod_{j \in \Gamma} \varphi(f_j(\lambda x))$$

$$+ \psi(\lambda x) \sum_{k \in \Gamma} \left(\prod_{j \neq k} \varphi(f_j(\lambda x)) \right) \varphi'(f_k(\lambda x)) f_k(\lambda^2 x)$$

$$< 0,$$

and the subclaim is proved.

End of the prof of the claim and the theorem. Using the implicit function theorem [233, XVI, 2.1] there is a uniquely determined function $g: U \to V$ of class C^k such that $F(x, g(x)) = 0$ for every $x \in U$. Clearly $g = \| \ \|_W$. □

Corollary 5.3.c. *For every set* Γ *the space* $c_0(\Gamma)$ *admits an equivalent* C^∞*-smooth norm.*

5.4   Rough norms and Asplund spaces

The classical definition of Asplundness is "every continuous convex function on X is Fréchet differentiable on some G_δ subset of X" It was considered in [331] as equivalent to "separable subspaces have separable duals" and proved that it is a 3SP property.

A norm $\| \ \|$ in a Banach space X is said to be *rough* if there exists $\varepsilon > 0$ so that for all $x \in X$ one has

$$\lim \sup_{\|h\| \to 0} \frac{\|x+h\| + \|x-h\| - 2\|x\|}{\|h\|} \geq \varepsilon.$$

This means that there exist "wrinkles" of the unit sphere that cannot be "brushed out" uniformly by means of a linear isomorphism. Since (see [82, I.5.3])

Proposition. 5.4.a. *A Banach space admits an equivalent rough norm if and only if it is not Asplund.*

then one has

Theorem 5.4.b. *Admitting no equivalent rough norm is a 3SP property.*

5.5   Gâteaux smooth renorming

We start with a basic trick to obtain spaces with Gâteaux smooth renorming.

Lemma 5.5.a. *Separable Banach spaces and $c_0(\Gamma)$ admit uniformly Gâteaux smooth norms.*

Proof. The key argument is to show that

> *If X and Z are Banach spaces, Z admitting an equivalent uniformly Gâteaux smooth norm, and if there exists a dense range operator $T: Z \rightarrow X$ then X admits a uniformly Gâteaux smooth renorming.*

To see this, consider the norm on X^* given by

$$|f|^2 = \|f\|^2 + \|T^*f\|^2$$

This norm is weak* lower semicontinuous on X^* and thus it is a dual norm. It can be shown that the predual norm on X is uniformly Gâteaux smooth [82, II.6.8].

Now, if X is separable, it is easy to construct operators $l_2 \rightarrow X$ with dense range. Analogously, the canonical inclusion $l_2(\Gamma) \rightarrow c_0(\Gamma)$ has dense range.   □

Theorem 5.5.b. *Admitting an equivalent Gâteaux smooth norm or an equivalent uniformly Gâteaux smooth norm are not 3SP properties.*

Proof. The counterexample is the sequence

$$0 \rightarrow C[0, 1] \rightarrow D[0, 1] \rightarrow c_0([0, 1)) \rightarrow 0$$

Recall that $D[0, 1]$ is the space of all real functions that are left continuous on $[0, 1]$ and have right limits at each point endowed with the supremum norm. The quotient map $q: D[0, 1] \rightarrow c_0([0, 1])$ defined by

$$qf = (f(x) - f(x^+))_{x \in [0, 1)}$$

has *Ker q* $= C[0, 1]$. Applying lemma 5.5.a both spaces $C[0, 1]$ and $c_0([0, 1))$

admit uniformly Gâteaux smooth norms: the former is separable and the later is some $c_0(\Gamma)$. It will also be convenient to recall the representation of $D[0, 1]$ as the space $C(K)$ of continuous functions on the punctured interval $K=[0, 1]\times\{0, 1\}$ endowed with lexicographical order (see also [82]). It was proved by Talagrand [131] that any equivalent norm on $D[0, 1]$ fails to be Gâteaux differentiable at some point of $C[0, 1] = \{f \in C(K) : f(x, 0) = f(x, 1)$ for every $x \in [0, 1]\}$.

Claim. The space $D[0, 1]$ does not admit an equivalent Gâteaux smooth norm.

Proof of the claim. Observe that any $f \in C[0, 1]$ such that $|f(1)| < \|f\|_\infty$ necessarily attains its maximum at two different points $x=(t, 0)$ and $x^+=(t, 1)$ (for some $t \in [0, 1]$) of K:

Let $\| \ \|$ be a δ-equivalent norm on $D[0, 1]$; i.e., for all $g \in D[0, 1]$

$$\delta^{-1} \|g\|_\infty \leq \|g\| \leq \delta \|g\|_\infty,$$

and assume that $\| \ \|$ is Gâteaux smooth at every point of $C[0, 1]$. If we denote by $\| \ \|^*$ the dual norm on $(C[0, 1], \| \ \|)^*$ and by δ_p the Dirac measure concentrated at point p, the Bishop-Phelps theorem allows us to find a measure ν_p on $[0, 1]$ with $\|\nu_p\|^* \leq (5\delta)^{-1}$ and such that $\mu_p = \delta_p + \nu_p$ attains its norm at some norm one function $f \in C[0, 1]$; i.e., $\mu_p(f) = \|\mu_p\|$. In what follows λ_p denotes a Hahn-Banach extension of μ_p in $M(K)=C(K)^*$.

If F is a finite subset of $[0, 1]$ then we write

$$U(F) = \{p \in (0, 1) : |\lambda_p(\{z\})| > (5\delta)^{-1} \text{ for each } z \in F\}.$$

Since $U(\emptyset)=(0, 1)$ is uncountable and $U(F)=\emptyset$ if F is large enough, there is some finite set F that is maximal for the property "$U(F)$ uncountable." Let U denote $U(F)$ for this set. Let $B^* = Ball((C[0, 1], \| \ \|)^*)$ endowed with the weak* topology. The compact metric space $L = [0, 1]\times B^*$ has the Lindelöf property, which means that every open covering of L admits a countable subcovering [220, Thm. 1.15]. Therefore, if for every $p \in U$ we denote by $(V_n(p))_{n \in \mathbb{N}}$ a countable basis of neighborhoods of μ_p in B^* and by H the uncountable set $\{(p, \mu_p) : p \in U\}$ then given positive integers k, n the set

$$K_{k, n} = \{p \in U : \quad \text{at least one of the sets}$$

$$H \cap [(p, p+k^{-1})\times V_n(p)] \text{ or } H \cap [(p-k^{-1}, p)\times V_n(p)] \text{ is countable } \}$$

is countable.

So, $\bigcup_{k,n} K_{k,n}$ is also countable and there exist some $p \in U$ such that

$$p \notin \bigcup_{k,n} K_{k,n} \cup F.$$

In this way, the sets

$$H \cap [(p, p+\varepsilon) \times V_n(p)] \text{ and } H \cap [(p-\varepsilon, p) \times V_n(p)]$$

are uncountable for every $\varepsilon > 0$ and every n. Moreover, the set

$$\{p \in U : |\lambda_p(\{x\})| > (5\delta)^{-1} \}$$

is countable. Thus it is possible to find sequences $(x(n))$ and $(y(n))$ in $[0, 1]$ such that:

    i) $x(n) < x < y(n)$

    ii) $lim_{n \to \infty} x(n) = lim_{n \to \infty} y(n) = x$

    iii) weak * $-lim_{n \to \infty} \mu_{x(n)} = $ *weak* * $-lim_{n \to \infty} \mu_{y(n)} = \mu_x$

    iv) $|\lambda_{x(n)}(\{x\})| \le (5\delta)^{-1}$
         $|\lambda_{y(n)}(\{x\})| \le (5\delta)^{-1}$.

By construction of μ_x, there exists $f \in C[0, 1]$ such that $\| f \| = 1$ and $\mu_x(f) = 1$. Let λ_1 and λ_2 be accumulation points of $\{\lambda_{x(n)}\}$ and $\{\lambda_{y(n)}\}$, respectively. One sees that both λ_1 and λ_2 belong to B^* and that $\lambda_1(f) = \lambda_2(f) = 1$. Thus, to prove that $\| \;\; \|$ is not Gâteaux differentiable at f it is enough, by Smullyan test 5.2.c, to see that $\lambda_1 \ne \lambda_2$.

To this end, take a sequence of functions $h_m \in C[0, 1]$ such that

$$\chi_{[x,1]} \le h_m \le \chi_{[x-1/m,1]}.$$

One has

$$\lambda_{y(n)}(h_m) = \delta_{y(n)}(h_m) + \nu_{y(n)}(h_m) \ge \frac{4}{5} \delta^{-1}.$$

Hence, by Lebesgue's theorem

$$\lambda_{y(n)}(\chi_{[x,1]}) \ge \frac{4}{5} \delta^{-1}.$$

Moreover, since

$$|\lambda_{y(n)}(\{x\})| \le (5\delta)^{-1}$$

then

$$\lambda_{y(n)}(\chi_{[x^+,1]}) \ge \frac{3}{5} \delta^{-1}.$$

On the other hand

$$\lambda_{x(n)}(h_m) = \delta_{x(n)}(h_m) + \nu_{x(n)}(h_m) \le \delta_{x(n)}(h_m) + (5\,\delta)^{-1}.$$

Hence, by Lebesgue's theorem

$$\lambda_{x(n)}(\chi_{[x,\,1]}) \le (5\,\delta)^{-1}.$$

And since

$$|\lambda_{x(n)}(\{x\})| \le (5\,\delta)^{-1}$$

one obtains

$$\lambda_{x(n)}\!\left(\chi_{[x^+,\,1]}\right) \le \frac{2}{5}\,\delta^{-1}.$$

In this form

$$\left|\lambda_1\!\left(\chi_{[x^+,\,1]}\right)\right| \le \frac{2}{5}\,\delta^{-1}$$

$$\left|\lambda_2\!\left(\chi_{[x^+,\,1]}\right)\right| \ge \frac{3}{5}\,\delta^{-1}$$

and $\lambda_1 \ne \lambda_2$. $\square$

Coban and Kenderov [73] showed that the set of points of Gateaux differentiability of the *sup* norm of $D[0, 1]$ does not contain a dense G_δ.

5.6   Weak Asplund spaces

The *weak Asplund* property means that continuous convex functions on X are Gâteaux differentiable on some dense G_δ subset of X. Subspaces of weakly compactly generated spaces are weak Asplund spaces [13] and, thus, both $C[0, 1]$ and $c_0[0, 1]$ are weak Asplund spaces. Talagrand shows in [339] that, assuming Martin's axiom, $D[0, 1]$ is not a weak Asplund space, and he does so constructing an exotic convex continuous (in fact 1-lipschitzian) function φ on D such that the set $G(\varphi)$ of its points of Gâteaux differentiability is dense and of first Baire category. The result of Coban and Kenderov just mentioned shows that the *sup* norm can play the role of "exotic convex function." See [298], final remarks. Therefore, the sequence $0 \to C[0, 1] \to D[0, 1] \to c_0([0, 1]) \to 0$ yields

Theorem 5.6.a. *To be a weak Asplund space is not a 3SP property.*

It is even unknown whether $X \oplus \mathbb{R}$ is weak Asplund assumed that X is weak Asplund.

5.7   D-spaces

This section is off limits, although we think that, as an application of the push-out, it can be interesting. Let $f: [0, 1] \to X$ be a continuous function with values on a quasi-Banach space X. A *primitive* for f is a differentiable function $F: [0, 1] \to X$ so that $F'(t) = f(t)$ for $0 \le t \le 1$. Kalton defines in [211] a *D-space* as a quasi-Banach space where all continuous functions have a primitive. Continuous functions on Banach spaces are Riemann integrable and therefore Banach spaces are *D*-spaces. Popov [297] poses the question of whether L_p spaces $0 < p < 1$ are *D*-spaces. Kalton [211 and 205] proves:

Proposition 5.7.a. *If X has trivial dual then it is a D-space.*

Then, Kalton defines the *core* of X as the maximal subspace with trivial dual and proves

Proposition 5.7.b. *If core X=0 then X is a D-space if and only if it is a Banach space.*

This, and the fact that quotients of *D*-spaces are *D*-spaces, show that if X is a *D*-space then the quotient $X/ \, coreX$ is a Banach space. Kalton gives then a counterexample for the converse (i.e., a non *D*-space X such that $X/coreX$ is a Banach space) that actually proves:

Theorem 5.7.c. *To be a D-space is not a 3SP property.*

Proof. Let $0 \to \mathbb{R} \to E \to l_1 \to 0$ be a nontrivial exact sequence as in 1.6.g. Let now be a space L with trivial dual satisfying $\lim_{n \to \infty} a_n(L) \, (n \log n)^{-1} = 0$ (recall from 1.5 that $a_n(X) = \sup \{ \, \| x_1 + \ldots + x_n \| : \| x_n \| \le 1 \}$). This estimate is relevant since [211, thm. 5] shows that in a *D*-space X one has the estimate $a_n(X) \le C \, a_n(core \, X)$ where C is some constant. It is not difficult to check that the space E satisfies the estimate $a_n(E) \ge k \, n \log n$ for some positive constant k. Let Σ be the push-out of the maps $\mathbb{R} \to E$ and some isometry $\mathbb{R} \to L$. Consider the lower exact row in the push-out diagram

$$
\begin{array}{ccccccccc}
0 & \to & \mathbb{R} & \to & E & \to & l_1 & \to & 0 \\
  &   & \downarrow &   & \downarrow &   & \| & & \\
0 & \to & L & \to & \Sigma & \to & l_1 & \to & 0
\end{array}
$$

Since Σ admits E as a subspace, $a_n(\Sigma) \ge a_n(E) \ge k \, n \log n$. But *core* Σ is isomorphic to L so that, were Σ a *D*-space, one should have $a_n(\Sigma) \le C \, a_n(L)$ which is impossible.   □

5.8   w*-G$_\delta$ extreme points

A related property was considered by Larman and Phelps in [234] suggested by Stegall's proof that weak Asplund spaces have weak* sequentially compact dual ball. They call a point $p \in X^*$ a w*-G$_\delta$ *extreme point* of a weak* compact set K if p is an extreme point of K and such that $\{p\}$ is a G_δ of K. For instance, weak* exposed points of K are w*-G$_\delta$ extreme points of K, but not conversely. The property w*-G$_\delta$ *extreme points exist* means that every nonempty weak* compact subset of X have at least one. Larman and Phelps prove:

Proposition 5.8.a. *If X^* admits w*-G$_\delta$ extreme points then it has weak* sequentially compact unit ball.*

We are however interested in:

Theorem 5.8.b. *To admit w*-G$_\delta$ extreme points is a 3SP property.*

Proof. Let $0 \to Y \to X \to Z \to 0$ be a short exact sequence where w*-G$_\delta$ extreme points exist in Y^* and Z^*. Suppose that K is a nonempty weak* compact subset of X^*. By hypothesis $i^*(K)$ has some w*-G$_\delta$ extreme point p and thus p is an extreme point of K such that for some sequence of weak* open sets (V_n) one has

$$\{p\} = \cap_n(V_n \cap i^*(K)).$$

Let $W = (i^*)^{-1}(p) \cap K$. This is a nonempty weak* closed extremal face of K. Choose $w \in W$ and note that $W - w \subset Y^\perp = Z^*$. Thus, there exists some w*-G$_\delta$ extreme point z for $W - w$ in Z^*; i.e., for some sequence (U_n) of weak* open subsets of Z^*

$$\{z\} = \cap_n(U_n \cap (W - w)).$$

Choose weak* open subsets (W_n) of Y^* such that, for each n, $U_n = W_n \cap Y^\perp$ and let $x = w + q$. Clearly x is an extremal point of W and thus of K. It is not difficult to verify that

$$\{x\} = \cap_n [(W_n + w) \cap (i^*)^{-1}V_n \cap K]$$

and the proof is complete.   □

5.9   Measure compact spaces

A Banach space X is said to be *measure compact* if all scalarly measurable X-valued functions φ are weakly equivalent to a Bochner measurable function; recall that a function $\psi: \Omega \to X$, $\mu(\Omega) = 1$, is said to be Bochner measurable if it is the pointwise limit *a.e.* of a sequence of measurable simple functions; equivalently, if it is scalarly measurable and there is a separable subspace $X_1 \subset X$ such that $\mu\{t \in \Omega: \psi(t) \notin X_1\} = 0$, and that two scalarly measurable functions $\varphi: \Omega \to X$ and $\psi: \Omega \to X$ are weakly equivalent if, for all $x^* \in X^*$, $x^*(\varphi) = x^*(\psi)$ *a.e.* Subspaces of weakly compactly generated spaces are measure compact spaces [106]; therefore $C[0, 1]$ and $c_0[0, 1]$ are measure compact spaces.

Theorem 5.9.a. *To be a measurable compact space is not a 3SP property.*

Proof. Consider again the sequence

$$0 \to C[0, 1] \to D[0, 1] \to c_0([0, 1]) \to 0$$

It turns out that the function $\varphi: [0, 1] \to D[0, 1]$ defined by $\varphi(t) = \chi_{[0, t]}$ is scalarly measurable and not weakly equivalent to any Bochner measurable function. The first assertion is easy, regarding $D[0, 1]$ as a subspace of $L_\infty [0, 1]$, since every element $x^* \in L_\infty[0, 1]^*$ is the difference $a - b$ of two positive functionals and $a(\varphi)$ (resp. $b(\varphi)$) is increasing, and thus measurable. Moreover $L_\infty[0, 1]^*$ is weak*-separable. This implies that two scalarly measurable functions are weakly equivalent if and only if they coincide almost everywhere (Hahn-Banach separation theorem). Now, if $A \subset [0, 1]$ is countable, $\varphi(A)$ is discrete in $L_\infty [0, 1]$ and hence norm separable. Thus, no Bochner measurable function weakly equivalent to φ can exist. □

5.10 Uniformly convex and locally uniformly convex renorming

A norm $\| \ \|$ is said to be *uniformly convex* if whenever x_n, $y_n \in S_X$ and

$$\lim_{n \to \infty} \|x_n + y_n\| = 0 \quad \text{then} \quad \lim_{n \to \infty} \|x_n - y_n\| = 0.$$

The following remarkable result is due to Enflo [114].

Proposition 5.10.a. *A Banach space admits a uniformly convex norm if and only if it is super-reflexive.*

Therefore,

Theorem 5.10.b. *Admitting an equivalent uniformly convex norm is a 3SP property.*

The norm $\| \ \|$ is said to be *locally uniformly convex* or *locally uniformly rotund* (in short *LUR*) if for all $x \in X$ whenever $\| x_n \| = \| x \|$ and

$lim_{n\to\infty} \| x_n + x \| = 2 \| x \|$ one has

$$lim_{n\to\infty} \| x_n - x \| = 0.$$

The following simple fact shall be useful (see [82, II.2.3]).

Proposition 5.10.c. *The norm* $\| \; \|$ *is LUR at* $x \in X$ *if and only if given* (x_n) *in* X *such that*

$$lim_{n\to\infty} 2 \| x \|^2 + 2 \| x_n \|^2 - \| x + x_n \|^2 = 0$$

one has

$$lim_{n\to\infty} \| x - x_n \|^2 = 0.$$

LUR norms on subspaces can be extended maintaining some properties:

Extension of *LUR* norms 5.10.d. (see [82, II.8.1]) *Let* Y *be a subspace of a Banach space* X. *An equivalent LUR norm* $| \; |$ *on* Y *can be extended to an equivalent norm on* X *that is LUR at every point of* Y.

Proof. If $| \; |$ is *LUR* on Y and we denote by $| \; |$ its extension to X given by 5.2.a then

$$[\![x]\!]^2 = |x|^2 + inf\{|x-y|^2 : y \in Y\}$$

defines an equivalent norm on X that is *LUR* at every point of Y. $\qquad\square$

Examples of spaces admitting *LUR* norms were provided by Troyanski [348]:

Lemma 5.10.e. *WCG spaces admit a LUR renorming.*

There are some duality relations (see [82, II.1.5 and II.1.6].

Proposition 5.10.f. *Let* $\| \; \|$ *be the norm of a Banach space* X, *and let* $\| \; \|^*$ *be the dual norm on* X^*. *If* $\| \; \|^*$ *is locally uniformly convex, then* $\| \; \|$ *is Fréchet smooth.*

Theorem 5.10.g. *To admit a LUR renorming is a 3SP property.*

Proof. Let Y be a subspace of a Banach space X. Assume that Y and X/Y admit a *LUR* renorming, and denote by $q: X \to X/Y$ the quotient map. We denote by $|.|_1$ the quotient norm on X/Y. By 5.10.d, there is an equivalent norm on X that is *LUR* at every point of Y.

Take a *LUR* equivalent norm $|.|$ on X/Y such that $|x+Y| \geq |x+Y|_1$ for every $x \in X$. Let $B: X/Y \to X$ be a continuous selection for q. Denote by S the unit sphere in $(X/Y, | \; |)$. For every $a \in S$ it is possible to take $f_a \in X^*$ such that:

$$f_a(Ba) = 1, \quad \|f_a\| = 1/|a|_1, \quad \text{and} \quad f_a|_Y = 0.$$

For $k \in \mathbb{N}$ and $a \in S$ let us define the maps P_a and $\Phi_{a,k}$ on X by

$$P_a(x) = f_a(x)\, Ba$$

$$\Phi_{k,a}(x) = |q(x)+a|^2 + k^{-1}\|x - P_a(x)\|^2(1+\|P_a\|)^{-2}.$$

Furthermore, for $x \in X$ let

$$\Phi_a(x) = sup\,\{\Phi_{k,a}(x) : a \in S\}$$

$$\Phi(x) = \|x\|^2 + |q(x)|^2 + \sum_{k=1}^{\infty} 2^{-k}\Phi_k(x).$$

Finally, let $\| \; \|_n$ be the Minkowski functional of the set

$$\{x \in X: \; \Phi(x) + \Phi(-x) \le 4\}.$$

It easy to see that $\| \; \|$ is an equivalent norm on X and it remains to show that it defines a *LUR* norm. This rather long and complicated proof can be seen in the original paper [136]. The idea is to apply the part corresponding to *LUR* points of the following result.

Lemma 5.10.h. *A vector $x \in X$ is a LUR point (resp. a point of weak-norm continuity, a denting point) for $\| \; \|_n$ in X provided qx is a LUR point for $|\;|$ in X/Y and $x - Bqx$ is a LUR point (resp. a point of weak-norm continuity, a denting point) for $\| \; \|$ in Y.*

The proof of the *LUR* case is in [136]. The proofs of the other cases appear in [189, Prop. 3.12]. □

A previous partial result in this direction was obtained by John and Zizler, *if X/Y is separable, and Y admits an equivalent LUR norm, then also X admits an equivalent LUR norm.*

5.11 *LUR* renorming
and fragmentability by slices

Lex X be a Banach space. Given $A \subset X, f \in X^*$ and $c \in \mathbb{R}$ we denote by

$$S(A, f, c) = \{x \in A: f(x) > c\},$$

the *open slice of A determined by f and c*. Given $\varepsilon > 0$, we say following [262] that

X has the *ε-sJNR property* if it admits a countable cover by sets that are union of slices of diameter less than ε; i.e., if we can write $X = \bigcup_{n=1}^{\infty} H_n$ so that for every $x \in H_n$ there is an open slice $S(H_n, f, c)$ of H_n which contains x and has diameter smaller than ε. We say that X has the *sJNR property* if it has the *ε-sJNR* property for every $\varepsilon > 0$.

Moltó, Orihuela and Troyanski proved in [262, Main Theorem]:

Proposition 5.11.a. *A Banach space admits an equivalent LUR norm if and only if it has the sJNR.*

Moreover, these authors showed a Decomposition Method [262, Proposition 11] from which it follows that *sJNR* is a 3SP property. The proof of the 3SP property we give here is essentially the proof of their Decomposition Method. We need the following Lemma.

Lemma 5.11.b. *If the norm of a Banach space Z is LUR, then for every $\varepsilon > 0$ it is possible to write $Z = \bigcup_{n=1}^{\infty} E_{\varepsilon}^n$ so that $E_{\varepsilon}^n \subset E_{\varepsilon}^{n+1}$ and for every $z \in E_{\varepsilon}^n$ there exists a norm one functional $g_z \in Z^*$ such that $g_z(z) = \|z\|$ and the following implication holds: if $w \in Z$ verifies*

$$\|w\| - 1/n < \|z\| < g_z(w) + 1/n$$

then

$$\|z - w\| < \varepsilon.$$

Proof. Fix $\varepsilon > 0$. Since the norm is *LUR*, for every non-zero $z \in Z$ we can find a norm-one $g_z \in Z^*$ satisfying $g_z(z) = \|z\|$ and $c(z) < 1$ so that

$$diam\ S(B_Z, g_z, c(z)) < \varepsilon/(2\|z\|).$$

For $n^{-1} < 2\varepsilon$ we set

$$E_{\varepsilon}^n = \{z \in Z:\ c(z) < (\|z\| + 1/n)^{-1}(\|z\| - 1/n)\ \}.$$

Now, if $z \in E_{\varepsilon}^n$ and $w \in Z$ satisfies $\|w\| - 1/n < \|z\| < g_z(w) + 1/n$ then $g_z(w/\|w\|) > (\|z\| - 1/n)/\|w\| > c(z)$. Thus $\|(z/\|z\|) - (w/\|w\|)\| < \varepsilon/(2\|z\|)$, hence $\|z - w\| < \varepsilon/2 + |\|w\| - \|z\|| < \varepsilon/2 + 1/n < \varepsilon$. $\square$

Theorem 5.11.c. *To have the sJNR is a 3SP property.*

Proof. Assume that Y is a subspace of X such that Y and $Z = X/Y$ have the *sJNR* property. We denote, as usual, by $q: X \to Z$ and $B: Z \to X$ the quotient map and a continuous selection for q, respectively; and by $P: X \to Y$ the map defined by $P(x) = z - Bqx$. It is enough to show that for a fixed $\varepsilon > 0$, the space X has the 2ε-*sJNR* property.

Since Y has the ε-*sJNR* property, we can write $Y = \bigcup_{n=1}^{\infty} H_n$ in such a way that for every $v \in H_n$ there exists a norm-one functional $h_v \in X^*$ and a real number μ_v so that the slice $S(H_n, h_v, \mu_v)$ contains v and has diameter less than ε. For every $j \in \mathbb{N}$ we write

$$H_n^j = \{v \in H_n: \, min\{h_v(v) - \mu_v, \, \varepsilon\text{-}diamS(H_n, h_v, \mu_v)\} > 2/j \}.$$

Clearly one has

$$Y = \bigcup_{n,j=1}^{\infty} H_n^j.$$

On the other hand, by Proposition 5.2.b, we can assume that the norm in Z is *LUR*. So we can consider the decomposition $Z = \bigcup_{n=1}^{\infty} E_{\varepsilon}^n$ given in Lema 5.11.b. Moreover, by the norm continuity of B, for every $j \in \mathbb{N}$ and every $z \in Z$ there exists $\eta^j(z) > 0$ such that $\|z - z'\| < \eta^j(z)$ implies $\|Bz - Bz'\| < 1/j$. Denoting

$$G_m^j = \{z \in Z: \, \eta^j(z) > 1/m\},$$

we have, for every $j \in \mathbb{N}$,

$$Z = \bigcup_{m=1}^{\infty} G_m^j.$$

Finally, we consider the subsets F_{ij}^k of X given by

$$F_{i,j}^k = \{x \in X: \, \|x\| < i/3, \, (k-1)(i^2 j) \le \|qx\| < k/(i^2 j)\},$$

and define

$$L_{kmn}^{ij} = P^{-1}(H_n^j) \cap q^{-1}(G_m^j \cap E_{\varepsilon}^n) \cap F_{ij}^k.$$

In this way we have

$$X = \bigcup_{i,j,k,m,n \in \mathbb{N}} L_{kmn}^{ij},$$

and this decomposition shows that X has the 2ε-*sJNR* property. Indeed, let us fix $w \in L_{kmn}^{ij}$. We denote $z = qw \in Z$, $u = Bqw$ and $v = Px \in Y$ (note that $w = u + v$). Take the functional $f \in X^*$ defined by $f(x) = g_z(qx) + i^{-2}h_v(x)$, and select a real number r so that

$$\mu_v + (2j)^{-1} < r < h_v(v) < r + 1/3.$$

One has

$$f(w) = \|z\| + i^{-2}h_v(u+v) > \|z\| + i^{-2}(h_v(u) + r).$$

We will show that if $\lambda = \|z\| + i^{-2}(h_v(u) + r)$ then the diameter of the slice $S(L_{kmn}^{ij}, f, \lambda)$ is smaller that 2ε. To do that, pick $y \in L_{kmn}^{ij}$. We have

$$g_z(qy) = f(y) - i^{-2}h_v(y)$$
$$> \lambda - i^{-2}h_v(y)$$
$$= \|z\| + i^{-2}(h_v(u) + r - h_v(y))$$
$$> \|z\| + i^{-2}(h_v(w - y) - \frac{1}{3})$$
$$> \|z\| + \frac{1}{i} .$$

Moreover, since y, $w \in F_{ij}^k$, we have that $\|qy\| > \|z\| - 1/i$; hence, we conclude (see lemma 5.11.b) that $\|qy - z\| < 1/m$ and $\|Bqy - u\| < 1/j$ because $z \in G_m^j$. On the other hand, we have

$$h_v(Py) = h_v(y) - h_v(Bqy)$$

$$= i^2(f(y) - g_z(qy)) - h_v(Bqy)$$

$$> i^2(\lambda - \|qy\|) - h_v(Bqy)$$

$$= i^2(\|z\| - \|qy\|) + r + h_v(u - Bqy)$$

$$> -\frac{i^2}{i^2 j} + r - \|u - Bqy\|$$

$$> r - \frac{2}{j}$$

$$> \mu_v .$$

Then we have $Py \in S(H_n^j, h_v, \mu_v)$, and it follows that $\|Pw - Py\| < \varepsilon - 2/j$; hence

$$\|w - y\| \leq \|Bqw - Bqy\| + \|Pw - Py\| < \varepsilon,$$

and the proof is finished. $\square$

5.12   Mazur intersection property

The norm of a Banach space X has the *Mazur intersection property* (*MIP*, in short) if every bounded closed convex subset of X can be obtained as an intersection of balls. A dual norm on X^* has the *weak*-MIP* if every bounded weak*-closed convex subset of X^* can be obtained as an intersection of balls. Mazur [259] was the first to study the *MIP* as an extension of a well-known property of the euclidean space. Giles, Gregory and Sims gave in [131] characterizations of the *MIP* and the *weak-MIP*.

Recall that a point p in the unit ball of X (resp. $p^* \in X^*$) is a *denting point* (resp. *weak*-denting point*) if for every $\varepsilon > 0$ there exist a functional $f \in X^*$ (resp. $z \in X$) and a number c so that $f(p) > c$ (resp. $p^*(z) > c$) and

$$\operatorname{diam} \{x \in Ball(X): f(x) > c\} < \varepsilon$$

$$(\text{resp. } \operatorname{diam} \{x^* \in Ball(X^*): x^*(z) > c\} < \varepsilon).$$

Proposition 5.12.a. [131] *The norm of a Banach space has the MIP if and only if the weak*-denting points of X^* form a dense subset of its unit sphere. The norm in X^* has the weak*-MIP if and only if the denting points of X form a dense subset of its unit sphere.*

In [131, Problem 2.2] it was asked whether spaces admitting a norm with *MIP* are Asplund. This question was open for some years, but recently Jiménez and Moreno [189] have given a strong negative answer, showing that every Banach space embeds isometrically into a Banach space with *MIP*. Moreover, these authors have obtained a partial 3*SP* result for *MIP* renorming; and using similar techniques they derived that admitting an equivalent norm whose dual norm is *LUR*, or has the weak*-*MIP*, are 3*SP* properties. Let us show these results.

Lemma 5.12.b. *A point $f \in X^*$ is a weak*-denting point if the norm of X^* is LUR at f.*

Proof. For every $\varepsilon > 0$ there exists $c > 0$ so that $g \in X^*$ and $1 - c < \| g \| \leq 1$ imply $\| f - g \| < \varepsilon$. Taking $x \in X$, $\| x \| = 1$ such that $f(x) > 1 - c$ one obtains that the diameter of the slice $\{g \in Ball(X^*): g(x) > 1 - c\}$ is smaller than ε. □

With this in combination with 5.12.a one has a partial 3*SP* result.

Proposition 5.12.c. *If X has a subspace Y admitting a norm whose dual norm has a dense set of LUR points and such that X/Y admits a MIP-norm then X admits a MIP-norm.*

Proof. The proof is based on the same construction used for *LUR* renorming in 5.10.g, but applied to $Y^\perp$ as a subspace of X^*. Using a norm in Y whose dual norm $|\ |$ has a dense set of *LUR* points, and considering in X^* a dual norm $\| \ \|$ obtained by extending the dual norm of $Y^\perp$ to a *MIP* norm in X/Y, Jiménez and Moreno construct a dual norm N in X^* given by the Minkowski functional of a certain set

$$\{f \in X^* : \Psi(f) + \Psi(-f) \leq 4 \}.$$

If b is a continuous selection for the quotient map $i^*\colon X^* \to Y^*$, they prove:

Lemma 5.12.d. *A point $f \in X^*$ is a LUR point (resp. a weak*-to-norm point of continuity; a weak*-denting point) for N in X^* provided i^*f is a LUR point for $|\ |$ in Y^* and $f - bi^*f$ is a LUR point (resp. a weak*-to-norm point of continuity; a weak*-denting point) for $\|\ \ \|$ in $Y^\perp$.*

The proof finishes taking into account that *LUR* points in the unit sphere of X^* are weak*-denting and applying proposition 5.12.a and lemma 5.12.d.   □

We pass to positive 3*SP* results.

Theorem 5.12.e. *MIP^{dual} is a 3SP property.*

Proof. Observe that a dual space with the *MIP* is reflexive [131, Corol. 2.7].   □

Theorem 5.12.f. *That X^* admits a LUR dual norm is a 3SP property.*

Proof. The proof goes as 5.10.g using the corresponding part of 5.12.d.   □

As Jiménez and Moreno realized, a space admitting a norm whose dual is *LUR* is not the same that a space whose dual space admits a *LUR* norm. So, 5.12.f is not the 3*SP* property for *LUR^{dual}*. However, since admitting a *LUR* renorming is a 3*SP* property,

Theorem 5.12.g. *LUR^{dual} is a 3SP property.*

Theorem 5.12.h. *That X^* admits a weak*-MIP dual norm is a 3SP property.*

Proof. It was proved by Troyanski (see [82], Chapter IV) that if D is the set of all denting points of the unit ball of X then X admits a norm that is *LUR* at each point of D. Then, by the characterization of a dual norm with weak*-*MIP* in Proposition 5.12.a, there exist equivalent norms in Y and X/Y with a dense set of *LUR* points in the sphere. To finish the proof it is enough to apply Lemma 5.10.h.   □

5.13   Szlenk index

Along with his study of the existence of *LUR* renormings of Banach spaces, Lancien [229] considered some ordinal indices associated to a Banach space.

Dentability index. Let X be a Banach space. For every $\varepsilon > 0$ we apply a slicing process to the unit ball of X as follows. Given a subset C of X, a *weak slice* of C is a subset of the form $\{x \in C : f(x) > \lambda\}$ for some non-zero element $f \in X^*$ and $\lambda \in \mathbb{R}$. Let us denote

$$C^1(\varepsilon) = \{x \in C : \text{weak-slices of } C \text{ containing } x \text{ have diameter } > \varepsilon\}.$$

By transfinite induction, if $\alpha+1$ is not a limit ordinal $C^{\alpha+1}(\varepsilon)=(C^\alpha)^1(\varepsilon)$. If α is a limit ordinal then $C^\alpha(\varepsilon) = \cap_{\beta<\alpha} C^\beta(\varepsilon)$.

The *dentability index* of C for an ε given is the smallest ordinal α for which $C^\alpha(\varepsilon)=C^{\alpha+1}(\varepsilon)$ (if it exists; say ∞ if not). We set $\delta(X, \varepsilon)$ to mean the dentability index of its unit ball corresponding to ε. The dentability index of X is defined by $\delta(X)= sup_{\varepsilon>0}\delta(X, \varepsilon)$. Lancien [230, Thm 1.2] proves (see also [231, remark p.639 and Thm. 2.1]):

Proposition 5.13.a.

   i) $\delta(X) < \omega$ *if and only if X admits a uniformly convex renorming.*
   ii) If $\delta(X) < \omega_1$ then X admits a LUR renorming.

Szlenk index. The same process applied to the dual space provides the Szlenk index. Its name comes from a similar index introduced by Szlenk [337] to prove that there is no universal separable reflexive space. For every $\varepsilon>0$ we apply a slicing process to the unit ball of X^* as follows. Given a subset C of X^*, a weak* slice of C is a subset of the form $\{f \in C : f(x)>\lambda\}$ for some non-zero element $x \in X$ and $\lambda \in \mathbb{R}$. Let us denote

$$C^1(\varepsilon) = \{f \in C : \text{weak* slices of } C \text{ containing } f \text{ have diameter } > \varepsilon\}.$$

By transfinite induction, if $\alpha+1$ is not a limit ordinal $C^{\alpha+1}(\varepsilon)=(C^\alpha)^1(\varepsilon)$. If α is a limit ordinal then $C^\alpha(\varepsilon) = \cap_{\beta<\alpha} C^\beta(\varepsilon)$.

The *Szlenk index* of C for an ε given is the smallest ordinal α for which $C^\alpha(\varepsilon)=C^{\alpha+1}(\varepsilon)$ (if it exists; say ∞ if not). We set $Sz(X, \varepsilon)$ to mean the Szlenk index of its dual unit ball corresponding to ε. The Szlenk index of X is defined by $Sz(X)= sup_{\varepsilon>0}Sz(X, \varepsilon)$. Lancien [230, Thm 1.2] proves:

Proposition 5.13.b. *If $Sz(X) < \omega_1$ then X admits an equivalent renorming whose dual norm is LUR.*

Moreover,

Theorem 5.13.c. *Having countable Szlenk index is a 3SP property.*

Proof. Precisely, if $0 \rightarrow Y \rightarrow X \rightarrow X/Y \rightarrow 0$ is a short exact sequence, Lancien proves a typical 3SP inequality:

$$Sz(X) \leq Sz(Y) \, Sz(X/Y).$$

This would be immediate consequence of

$$i^* \left[Ball(X^*)^{\,\omega\, Sz\,(X/Y,\,\frac{\varepsilon}{9})\,\alpha}\,(\varepsilon) \right] \subset Ball(Y^*)^\alpha \left(\frac{\varepsilon}{4}\right).$$

and a technical result on ordinals: if $\gamma < \omega^\alpha$ then $\omega\gamma\omega^\beta \leq \omega^\alpha\omega^\beta$. The proof of the containment needs the following result.

If Z^* is separable then

$$\left[3Ball(Y^\perp) + \frac{\varepsilon}{3} Ball(X^*) \right]^{\omega\,\alpha} (\varepsilon) \subset \left(3\,Ball(Y^\perp)\right)^{\omega\,\alpha}\left(\frac{\varepsilon}{3}\right) + \frac{\varepsilon}{3} Ball(X^*),$$

which can be obtained by transfinite induction. The reduction from the general case to "Z^* separable" can be done thanks to the following lemma which is easy to state but cumbersome to prove (see [230]).

Lemma 5.13.d. *Let $\varepsilon > 0$ and X a Banach space. If $Sz(X, \varepsilon) < \omega_1$ then there is a closed separable subspace Y of X such that $Sz(Y, \varepsilon/2) = Sz(X, \varepsilon)$.*

With this, let M be a separable subspace of X and let E be the closed span of $M \cup Y$. If $(E/Y)^*$ is separable, it has already been proved that

$$Sz(E) \leq Sz(Y)Sz(E/Y) \leq Sz(Y)Sz(X/Y).$$

Thus

$$Sz(M) \leq Sz(E) \leq Sz(Y)Sz(X/Y),$$

and the lemma 5.13.d yield the desired inequality.   □

Since it was already proved that *LUR* renorming is a *3SP* property, but it is unknown whether it is equivalent to "having countable dentability index", one might ask

Question. *Is "having countable dentability index" a 3SP property?*

5.14   Midpoint *LUR* renorming

A norm $\| \ \|$ is *midpoint locally uniformly convex (MLUR)* if $x \in S_X$ and

$$\lim_{n\to\infty}(\|x+y_n\| + \|x-y_n\|) = 0$$

imply $\lim_{n\to\infty} y_n = 0$.

LUR norms are *MLUR* norms. However, Haydon [165] has shown that there exist spaces with *MLUR* norm not admitting equivalent *LUR* norm. The example of Haydon [165] displayed in 5.17.d serves here since strictly convex norms are *MLUR* norms.

Theorem 5.14.a. *MLUR renorming is not a 3SP property.*

Partial 3*SP* results appear in [3] and also [82, p. 332]; for instance, *If Y admits a MLUR norm and X/Y admits a LUR norm, then X admits a MLUR norm.*

5.15   The Krein-Milman property

A Banach space X has the *Krein-Milman* property (*KMP*) if every closed bounded convex subset of X is the closed convex hull of its extreme points. The *RNP* implies *KMP* [93, VII 1.7]. The RNP and *KMP* are equivalent in dual spaces [93, VII 2.8], in Banach lattices [40], and in Banach space isomorphic to their square [320]. It is an open question explicitly posed by in [93, p. 209] whether they are equivalent.

It is unknown if *KMP* is even stable by products.

5.16   Asymptotic norming properties

In [177], James and Ho introduced three kinds of so-called *asymptotic norming properties*: *ANP-k* ($k=$I, II and III). They proved that *ANP*-I implies *KMP*; that *ANP*-III implies *RNP*; and that for separable spaces the three properties coincide. Ghoussoub and Maurey [129] proved that for separable spaces *ANP* and *RNP* coincide. The 3*SP* character of the *ANP* has been studied by Hu and Lin [172]. Using techniques developed to prove the 3*SP* property for *LUR* renorming, they showed that *ANP*-I is a 3*SP* property, and obtained partial results for the other two.

A *norming set* Φ for a Banach space X is a subset of the unit ball of X^* such that for every element $x \in X$,

$$\| x \| = sup \, \{ f(x) : f \in \Phi \}.$$

A sequence (x_n) is said to be *asymptotically normed* by a subset $\Phi \subset Ball(X^*)$ if for every $\varepsilon > 0$ there are a functional $f \in \Phi$ and an integer m so that

$$f(x_n) > \| x_n \| - \varepsilon \quad \text{for all } n > m.$$

A sequence (x_n) in a Banach space X is said of type k if:

I. (x_n) is convergent

II. (x_n) has a convergent subsequence

II'. (x_n) is weakly convergent

III. The intersection $\bigcap_n conv(x_n, x_{n+1}...)$ is non-empty.

A Banach space X is said to have ANP-k (k=I, II or III) if it has an equivalent norm for which there is a norming set Φ so that every normalized sequence that is asymptotically normed by Φ satisfies k.

Clearly ANP-$I \Rightarrow ANP$-$II \Rightarrow ANP$-III. It is easy to see that l_1 has ANP-II for its usual norm (just let Φ be the set of sequences with finite nonzero entries, all $+1$ or -1). The three properties are equivalent in separable spaces and coincide with the RNP [129].

Theorem 5.16.a. *Property ANP-I is a 3SP property.*

Proof. Let $0 \to Y \to X \to Z \to 0$ be a short exact sequence where Y and Z have ANP-I. Since ANP-I implies LUR renorming [170, Corol. 2.8] and 5.10.g, X admits a LUR renorming, and thus ANP-II. In [170, Thm. 2.5] it is proved that ANP-II in strictly convex spaces is equivalent to ANP-I $\qquad\square$

Partial positive results [172] are:

Proposition 5.16.b. *Let Y be a subspace of a Banach space X. Assume that X/Y has ANP-I. If Y has ANP-k (k=II,III) then X has ANP-k.*

Sketch of Proof. *Step 1.* *Embedding of X into some dual space Z^*.* Take circled subsets Φ_1 in $(X/Y)^* = Y^\perp \subset X^*$ and Φ_2 in Y^* given by the definition of the ANP properties. Lift Φ_2 to a bounded subset of X^* and consider the Minkowski functional $|\ |_3$ of $conv(\Phi_1 \cup \Phi_2)$. The completion of the subspace generated by $\Phi_1 \cup \Phi_2$ with respect to $|\ |_3$ is denoted Z.

The evaluation map $\delta: X \to Z^*$ given by $\delta x(z) = z(x)$ is an isomorphic embedding.

Step 2. *Suitable renorming of Z^*.* For $z^* \in Z^*$ define two (i=1, 2) continuous weak*-lower semicontinuous seminorms on Z^* by

$$|z^*|_i = sup_{z \in \Phi_i} z^*(z) .$$

Moreover, the predual norms satisfy

$$|x|_1 = \| qx \|_1 \text{ for all } x \in X;$$
$$|y|_2 = \| qy \|_2 \text{ for all } y \in Y.$$

Define now an equivalent dual norm in Z^* by

$$\|z^*\| = \left(\sum_{i=1}^{3} |z^*|_i^2\right)^{1/2}.$$

With the method of proof of 5.10.g it is possible to obtain a new equivalent norm on Z^* that, when restricted to X, has ANP-k for $k=$I or II.   □

Another partial result is also possible.

Proposition 5.16.c. *Let Y be a reflexive subspace of X. If X/Y has ANP-III, then X has ANP-III.*

Proof. Clearly, reflexive spaces have ANP-III; hence, since they admit LUR renorming, they have ANP-I.   □

Hu and Lin introduce in [171] the w^*-ANP-k properties of types $k=$I, II, II' or III: if every asymptotically normed sequence in the unit sphere of X^* has property k. These properties are studied in [18], where it is obtained the following result.

Theorem 5.16.d. w^*-ANP-k $(k = $ I, II, II'$)$ *are not 3SP properties.*

Proof. In fact, what one proves is that those properties are not stable by products in very much the way that was done with the Hahn-Banach smooth property at 1.9.h. The key is again to prove that a non-reflexive space with any of those properties has no nontrivial L-projections. On the other hand, the space $X\oplus_1 X$ admits nontrivial L-projections.   □

5.17   Strictly convex renorming

A norm $\|\ \|$ is said to be *strictly convex*, or *rotund*, if $x=y$ whenever $x, y \in X$ are such that $\|x\| = \|y\| = 1$ and $\|x+y\| = 2$.

Proposition 5.17.a. (see [82]) *Let $\|\ \|$ be the norm of a Banach space X, and let $\|.\|^*$ be the dual norm on X^*. If $\|\ \|^*$ is strictly convex then $\|\ \|$ is Gâteaux smooth. If $\|\ \|^*$ is Gâteaux smooth then $\|\ \|$ is strictly convex.*

Clarkson [27] proved that every separable Banach space admits a strictly convex renorming. This, and the fact that l_∞ has an equivalent strictly convex norm easily follow from

Proposition 5.17.b. *If X is a Banach space and there exists an injective operator $T: X \to Y$ into a strictly convex space then*

$$|x|^2 = \|x\|^2 + \|Tx\|^2$$

defines a strictly convex norm on X.

Bourgain [31] proved that l_∞/c_0 admits no strictly convex renorming.

Day's strictly convex norm on $c_0(\Gamma)$. Let Γ be an arbitrary set. Day's norm on $c_0(\Gamma)$ is defined by

$$\|x\|_D = \sup\left(\sum_{k=1}^n 4^{-k} x(\gamma_k)^2\right)^{1/2},$$

where the supremum is taken over all finite subsets $(\gamma_1,\ldots,\gamma_n)$ of distinct elements of Γ. That this norm is strictly convex can be seen as follows: assume that $\|x\| = \|y\| = 1$ and $\|x+y\| = 2$;

$$0 \le \sum 4^{-k}(x(\gamma_k)-y(\gamma_k))^2$$

$$= \sum 4^{-k}\left(2x(\gamma_k)^2+2y(\gamma_k)^2-(x(\gamma_k)+y(\gamma_k))^2\right)$$

$$\le 2\|x\|^2+2\|y\|^2-\|x+y\|^2$$

$$= 0.$$

Hence $x=y$.   □

With some extra work, it can be shown that Day's norm is *LUR*.

Given a locally compact space L we denote by $L^*=L\cup\{\infty\}$ the one-point compactification of L, and by $C_0(L)$ the space of continuous functions f on L^* such that $f(\infty)=0$. Observe that $C_0(L)$ is an hyperplane of $C(L^*)$.

In what follows a *tree* is a partially ordered set $(\Xi, \le)$ satisfying two properties: i) for every $t \in \Xi$ the set $\{s \in \Xi: s\le t\}$ is well ordered by $\le$; ii) every totally ordered subset of Ξ has at most one minimal upper bound. For convenience we introduce two elements 0 and ∞ that are not in Ξ and satisfy $0<t<\infty$ for all $t\in\Xi$. If $s, u \in \Xi$ then we denote $(s, u] = \{t \in \Xi : s<t\le u\}$. The *order* topology on a tree has as a basis of neighborhoods of t the intervals $(s, t]$ with $s<t$. The set of *immediate successors* of an element t shall be denoted t^+. We write Ξ^+ for the set of successor elements of Ξ; i.e., $\bigcup_{s\in\Xi\cup\{0\}} s^+$. If $t \in \Xi^+$ then there is a unique element t^- for which $(t^-)^+ = t$. We say that a tree is *dyadic* if for all elements t, t^+ has exactly two elements.

A function $\rho: \Xi \to \mathbb{R}$ is said to be strictly increasing if $s<t$ implies $\rho(s)<\rho(t)$. A tree Ξ is said to be $\mathbb{R}$-embeddable if there is some strictly increasing real valued function on it. The notion of $\mathbb{R}$-embeddability is relevant for us since

Proposition 5.17.c. *If Ξ is an $\mathbb{R}$-embeddable tree then $C_0(\Xi)$ admits a strictly*

convex renorming.

Proof. Assume that $\rho: \varXi \to \mathbb{R}$ is a strictly increasing function. It can be assumed without loss of generality that ρ takes values in $(0, 1]$ and we set $\rho(0)=0$. We define an operator $T: C_0(\varXi) \to l_\infty(\varXi)$ by

$$Tf(t) = \begin{cases} f(t)\left(\rho(t) - \rho(t^-)\right) & \text{if } t \in \varXi^+ \\ 0 & \text{otherwise.} \end{cases}$$

Clearly T is continuous with $\|T\| = \sup_{t \in \varXi} (\rho(t) - \rho(t^-)) \leq 1$. We shall show that T takes values in $c_0(\varXi)$. Since the linear span of the indicator functions $\mathbf{1}_{(0, t]}$, $t \in \varXi$ is dense in $C_0(\varXi)$, it is enough to prove that $T\mathbf{1}_{(0, 1]} \in l_1(\varXi) \subset c_0(\varXi)$ for all $t \in \varXi$. Given $t \in \varXi$ one has

$$\sum_{s \in \varXi} |T\mathbf{1}_{(0, t]}(s)| = \sum_{s \in \varXi^+ \cap (0, t]} \rho(s) - \rho(s^-) \leq \rho(t).$$

The operator T is injective: if $f \in Ker\ T$ then $f(t)=0$ for all $t \in \varXi^+$ and $f=0$ since $\varXi^+$ is dense in $\varXi$. Thus, $T: C_0(\varXi) \to c_0(\varXi)$ is an injective operator, and using Proposition 5.17.b and Day's strictly convex renorming, the result is proved. $\square$

Theorem 5.17.d. *Admitting an equivalent strictly convex norm is not a 3SP property.*

The setting. The construction starts with an $\mathbb{R}$-embeddable tree Λ. This tree Λ shall be embedded as a closed subtree of another tree Θ in such a way that $\Theta\backslash\Lambda$ is discrete. It turns out that both $Y=\{f \in C_0(\Theta) : f(\Lambda)=0\} = c_0(\Theta\backslash\Lambda)$ and $C_0(\Theta)/Y = C_0(\Lambda)$ admit strictly convex renormings. It remains to show how to achieve this construction to obtain that $C_0(\Theta)$ admits no strictly convex renorming.

Given a tree $\varXi$ and an increasing function $\rho: \varXi \to \mathbb{R}$ we shall say that a point $t \in \varXi$ is a *good point* for ρ if

$$card\ \{u \in t^+ : \rho(u)=\rho(t)\ \} < \infty.$$

Points that are not good points are called *bad points*. If $\|\ \|$ is an equivalent norm on $C_0(\varXi)$ then the increasing function

$$\mu(t) = inf\ \{\ \|\mathbf{1}_{(0, t)} + g\ \| : supp\ g \subset (t, \infty)\}$$

is called the associated function (to the new norm).

Lemma 5.17.e. *Let $\|\ \|$ be an equivalent renorming of $C_0(\varXi)$. If t is a bad point for the associated function μ then $\mu(t)= \|\mathbf{1}_{(0, t]}\|$.*

Proof. It is clear from the definition of μ that

$$\| \mathbf{1}_{(0,\,t]} \| \geq \mu(t).$$

"Badness" implies that there is a sequence (u_n) of distinct elements of t^+ such that $\mu(u_n) \to \mu(t)$. Thus, there are functions f_n having the form $f_n = \mathbf{1}_{(0,\,u_n]} + g_n$ with *supp* $g_n \subset (u_n, \infty)$ such that $\| f_n \| \to \mu(t)$. The sequence (f_n) being bounded and pointwise convergent in $C_0(T)$ to $\mathbf{1}_{(0,\,t]}$, converges weakly. Since the norm is weakly lower semicontinuous

$$\| \mathbf{1}_{(0,\,t]} \| \leq \lim\inf \| f_n \| = \mu(t). \qquad \square$$

Proposition 5.17.f. *Let T be a tree and assume that $C_0(T)$ admits an equivalent strictly convex norm $\| \ \|$. The associated function μ is strictly increasing in the set of its own bad points.*

Proof. Suppose that t and u are bad points with $t \leq u$ and $\mu(t) = \mu(u) = \alpha$. Then we have, by lemma 5.17.e

$$\| \mathbf{1}_{(0,\,t]} \| = \alpha = \| \mathbf{1}_{(0,\,u]} \|$$

while, by the definition of μ, one has

$$\| (1/2)(\mathbf{1}_{(0,\,t]} + \mathbf{1}_{(0,\,u]}) \| = \| \mathbf{1}_{(0,\,t]} + \mathbf{1}_{(t,\,u]} \| \geq \alpha.$$

The strict convexity of the norm implies $\mathbf{1}_{(0,\,t]} = \mathbf{1}_{(0,\,u]}$, which yields $t = u$. $\square$

Therefore, we need a tree Θ containing Λ as a closed subset with $\Theta \backslash \Lambda$ discrete and such that

(‡)   *No increasing function on Θ is strictly increasing on the set of its own bad points.*

Construction of the tree. Let Λ be the set of functions t having as domain a countable ordinal and as range a subset of ω having complementary infinite. This tree is $\mathbb{R}$-embeddable since the function $t \to \Sigma_{\alpha \in \text{dom } t} \, 2^{-t(\alpha)}$ is strictly increasing.

In [164, Prop 4] it is proved the following fact: *There exists a topology τ on the tree Λ so that (Λ, τ) is a Baire space.* Consequently,

If $\rho : \Lambda \to \mathbb{R}$ is increasing then there exists $t \in \Lambda$ such that for every $\varepsilon > 0$ the set $\{u \in t^+ : \rho(u) > \rho(t) + \varepsilon\}$ is finite.

Proof. The function ρ is τ-lower semicontinuous and, since it is defined on a Baire space, has points of τ-continuity. Any such point has the required property. $\square$

To construct Θ, for each $t \in \Lambda$ partition t^+ in two infinite disjoint subsets t_0 and t_1 and define $\Theta = \Lambda \times \{0, 1\}$ equipped with the order relation

$$(t, i) < (v, j) \Leftrightarrow v = t \text{ and } i = 0, j = 1 \quad \text{or} \quad \text{there exists } u \in t_i : u \leq v.$$

The subset $\Lambda \times \{0\}$ is homeomorphic to Λ and $\Lambda \times \{1\}$ is open and discrete. If we identify Λ with $\Lambda \times \{0\}$ in Θ, then we may regard Θ as having been obtained from Λ by introducing extra points $(t, 1)$ in such a way that $(t, 1)$ is between t and u for all those elements u that were immediate successors of t in Θ.

Proof of condition (‡) *and end of the proof of theorem 5.17.d.* Given and increasing function $f: \Theta \to \mathbb{R}$ we define $\rho: \Lambda \to \mathbb{R}$ by $\rho(t) = f(t, 0)$. Let $t \in \Lambda$ be one of those points whose existence was proved in the *claim*, i.e., a point such that for all $\varepsilon > 0$, $card \{ u \in t^+ : \rho(u) > \rho(t) + \varepsilon \} < \infty$. Let (u_n) be a sequence of distinct points of t_1. Since $(t, 0) \leq (t, 1) \leq (u_n, 0)$ in Θ,

$$\rho(t) = f(t, 0) \leq f(t, 1) \leq f(u_n, 0) = \rho(u_n)$$

for all n. By the choice of t, $\rho(u_n) \to \rho(t)$ as $n \to \infty$, which implies that $f(t, 1) = f(t, 0)$ and that $(t, 1)$ is a bad point for f.

Considering a sequence of distinct elements of t_0 it can be show similarly that $(t, 0)$ is a bad point for f. □

Following [82, VII.3.3] and modifying the proof of the 3*SP* property for *LUR* renorming, it may be proved that

Proposition 5.17.g. *If Y admits a strictly convex renorming and X/Y admits a LUR renorming then X admits a strictly convex renorming.*

5.18   λ-property

Given a normed space X, a triple (e, y, λ) is said to be *amenable to x* if e is an extreme point of *Ball(X)*, $y \in Ball(X)$, and $0 < \lambda \leq 1$ is such that $x = \lambda e + (1 - \lambda)y$. The λ-*function* is then defined as

$$\lambda(x) = sup \{\lambda : (e, y, \lambda) \text{ is amenable to } x\}.$$

If no triple (e, y, λ) is amenable to X then we set $\lambda(x) = 0$. The space X is said to have the λ-*property* if $\lambda(x) > 0$ for all $x \in Ball(X)$. If, in addition,

$$inf \{\lambda(x) : x \in Ball(X)\} > 0$$

then it is said that X has the *uniform* λ-*property*.

In a strictly convex space every point of the unit ball is a convex combination of two extreme points; thus, every strictly convex space has the uniform λ-property and this geometrical property can be viewed as a generalization of strict convexity. It was introduced in [11]. Other interesting features appear in [251]. In [252], Lohman proves

Theorem 5.18.a. *The uniform λ-property is not a 3SP property.*

Proof. As it is well known, the unit ball of c_0 has no extreme points. Let $\| \quad \|$ denote a strictly convex renorming of c_0 as in 5.17, and let V be its unit ball. Let X be the product $c_0 \times \mathbb{R}$ endowed with the norm having as unit ball the absolutely convex closed hull of

$$B = conv (V \times \{0\} \cup Ball(c_0) \times \{-1, 1\}).$$

It is simple to verify that

$$0 \to (c_0, \| \quad \|) \to X \to \mathbb{R} \to 0$$

is a short exact sequence. Thus, both the subspace and the quotient, being strictly convex, have the uniform λ-property.

To prove that X has not the uniform λ-property observe that the quotient map q is also a functional on X. Had X the uniform λ-property then q would attain its norm at some extreme point $(x, t) \in B$, see [11, Thm. 3.3]. Since $\| q \| = 1$, some point $(x, 1)$ is an extreme point of B. That implies $x \in Ball(c_0)$. But since $Ball(c_0)$ does not have extreme points, $(x, 1)$ cannot be an extreme point of B, and this is a contradiction. $\qquad\qquad\square$

Observe the close similitude between the construction of the unit ball B and that of the unit ball of a Snarked sum:

$$C = conv (\{(y, 0) : y \in Ball(Y) \} \cup \{(Fz, z) : z \in Ball(Z) \}).$$

5.19 Octahedral norms

The norm $\| \quad \|$ is said to be *octahedral* if for every finite-dimensional subspace F of X and every $\varepsilon > 0$ there exists $y \in S_X$ so that for every $x \in X$ one has

$$\| x + y \| \geq (1 - \epsilon)(\| x \| + 1) .$$

The following result is in [82, III.2.5].

Proposition. 5.19.a. *A Banach space admits an equivalent octahedral norm if and* *only if it contains a subspace isomorphic to l_1.*

Therefore,

Theorem 5.19.b. *Admitting no equivalent octahedral norm is a 3SP property.*

5.20   Kadec and Kadec-Klee norms

A norm on X is said to be a *Kadec norm* (resp. a *Kadec-Klee norm*) if weak convergence of nets (resp. sequences) on the unit sphere S_X is equivalent to norm convergence of nets (resp. sequences). A dual norm on X^* is said to be a *weak*-Kadec norm* (resp. *weak*-Kadec-Klee norm*) if weak* convergence of nets (resp. sequences) on the unit sphere is equivalent to norm convergence of nets (resp. sequences).

Proposition 5.20.a. [189]

(*a*) *If X/Y admits an equivalent LUR norm and Y admits an equivalent Kadec norm then X admits an equivalent Kadec norm.*

(*b*) *If Y admits an equivalent norm whose dual norm is LUR and $(X/Y)^*$ admits a dual weak*-Kadec norm then X^* admits a dual weak*-Kadec norm.*

Proof. (*a*) Observe that the norm of X is a Kadec norm if and only if every point in the unit sphere is a point of relative weak-to-norm continuity. So, using the construction in the proof of *LUR* renorming 5.10.g. and applying the corresponding part of lemma 5.10.h we get the result.

(*b*) The norm of X^* is a weak*-Kadec norm if and only if every point in the unit sphere of X^* is a point of weak*-to-norm continuity. So, it is enough to use the construction of 5.10.g and the corresponding part of lemma 5.12.d.   □

Proposition. 5.20.b. ([368] and [242]) *If X/Y admits an equivalent LUR norm and Y admits an equivalent Kadec-Klee norm then X admits an equivalent Kadec-Klee norm.*

Proof. The proof in [242] follows the same construction as in the proof of 5.10.g, and is similar to that of the previous results.   □

5.21   Ranges of vector measures

Not many things are known about the range of a X-valued countably additive vector measure μ. Lyapunov convexity theorem asserts that if X is finite dimensional and μ is nonatomic then the range of μ is convex [93; Corol. IX.1.5]. Nevertheless, Lyapunov convexity theorem fails in every infinite dimensional Banach space [93; IX.1]. This fact lead Kadets and Schechtman [199] to introduce the *Lyapunov property* of a Banach space X as: every nonatomic X-valued measure has the closure of its range convex. It is not difficult to see that

$L_2[0, 1]$ *fails the Lyapunov property.*

Proof. Consider the σ-algebra Σ of Borel subsets of $[0, 1]$ and the measure $\mu\colon \Sigma \to L_2[0, 1]$ defined by $\mu(A) = \mathbf{1}_A$. One has:

$$2^{-1}\,\mathbf{1}_{[0,\,1]} = 2^{-1}\,\mu(\varnothing) + 2^{-1}\,\mu([0,\,1]),$$

while for every $A \in \Sigma$

$$\|\,\mathbf{1}_A - 2^{-1}\,\mathbf{1}_{[0,\,1]}\,\|_2 = 2^{-1}. \qquad\qquad \square$$

However, the spaces l_p, $1 \le p < \infty$ and $p \ne 2$, satisfy the Lyapunov property [199]. In [200], Kadets and Vladimirskaya proved

Theorem 5.21.a. *The Lyapunov property is a 3SP property.*

The proof is based on the following characterization.

Proposition 5.21.b. *Let $\mu\colon \Sigma \to X$ be a nonatomic measure. The closure of $\mu(\Sigma)$ is convex if and only if for every $A \in \Sigma$ and $\varepsilon > 0$ there exists $B \in \Sigma$ such that $B \subset A$ and $\|\,\mu(B) - 2^{-1}\mu(A)\,\| < \varepsilon$.*

Proof. The "only if" part is clear. To prove the "if" implication, it is enough to prove that for every $x, y \in \mu(\Sigma)$ one has $2^{-1}(x+y) \in \overline{\mu(\Sigma)}$. So, suppose $x = \mu(A)$ and $y = \mu(B)$. Choose $U \subset A \backslash B$ and $V \subset B \backslash A$ in such a way that

$$\|\,\mu(U) - 2^{-1}\mu(A\backslash B)\,\| < \varepsilon$$

and

$$\|\,\mu(V) - 2^{-1}\mu(B\backslash A)\,\| < \varepsilon.$$

Let $W = U \cup V \cup (A \cap B)$. One has

$$\|\,\mu(W) - 2^{-1}(x+y)\,\| \;=\; \|\,\mu(U) - 2^{-1}\mu(A\backslash B) + \mu(V) - 2^{-1}\mu(B\backslash A)\,\| < 2\varepsilon,$$

and the proof is finished. $\square$

Proof of Theorem 5.21.a *(Sketch).* Let $0 \to Y \to X \to Z \to 0$ be a short exact sequence where Y and Z satisfy the Lyapunov property and let $\mu \colon \Sigma \to X$ be a nonatomic measure.

In order to apply the previous characterization, fix $A \in \Sigma$ such that $\mu(A) \neq 0$ and fix $\varepsilon > 0$. Moreover, select a probability measure λ on Σ such that $\lambda(A) \neq 0$. Denote by Σ_A the elements of Σ that are contained in A. we define $\nu \colon \Sigma_A \to X$ by

$$\nu(B) = \mu(B) - \lambda(B)\lambda(A)^{-1}\mu(A).$$

Since Z has the Lyapunov property, Kadets and Vladimirskaya prove [200, Lemmas 3 and 4] the existence of a σ-algebra $\Sigma' \subset \Sigma_A$ such that λ is nonatomic on Σ' and satisfies that for every $B \in \Sigma'$

$$\| q\nu(B) \| < 2^{-1}\varepsilon\lambda(B);$$

(here, as usual, $q \colon X \to Z$ denotes the natural quotient map).

Using this fact and [200, Lemma 6] it is possible to select a σ-algebra $\Sigma'' \subset \Sigma'$ and a nonatomic measure $\beta \colon \Sigma'' \to Y$ such that λ is nonatomic on Σ'' and for every $B \in \Sigma''$

$$\| \nu(B) - \beta(B) \| < 2\varepsilon.$$

Now, the Lyapunov property of Y allows one to find $B_0 \in \Sigma''$ such that $\lambda(B_0) = 2^{-1}\lambda(A)$ and

$$\| \beta(B_0) - 2^{-1}\beta(A) \| < \varepsilon.$$

Since $\nu(A) = 0$, one has

$$\| \mu(B_0) - 2^{-1}\beta(A) \|$$
$$= \| \nu(B_0) - 2^{-1}\nu(A) \|$$
$$\leq \| \nu(B_0) - \beta(B_0) \| + \| \beta(B_0) - 2^{-1}\beta(A) \| + \| 2^{-1}\beta(A) - 2^{-1}\nu(A) \|$$
$$< 4\varepsilon. \qquad \square$$

This result provides many new examples of spaces with Lyapunov's property, such as twisted sums of l_p spaces. Other examples of spaces with this property can be seen in [153].

A classical result is that the range of a vector valued countably additive vector measure is a weakly compact set. Diestel and Seifert proved in [91] that it has even the Banach-Saks property. In [9], Anantharaman and Diestel showed that weakly 2-summable sequences lie in the range of a vector measure and asked whether sequences in the range of a vector measure admit weakly 2-summable subsequences.

That this is not always the case was shown in [66]. It was actually proved that the range of a vector measure does not necessarily have the p-Banach-Saks property for some $1 < p$. However, as it follows from results of Rosenthal [312], reflexive quotients of a $C(K)$ space are superreflexive; from this (see [66]) it follows that if the unit ball of X lies inside the range of a vector measure then $X \in W_2$ (i.e., every bounded sequence in X admits a weakly 2-convergent subsequence). Since the space Z_2 has not property W_2, one has

Theorem 5.21.c. *To have the unit ball contained in the range of a countable additive vector measure is not a 3SP property.*

5.22   The U.M.D. property

A Banach space X is said to have the *U.M.D.* property if X-valued martingale difference sequences are unconditional in $L_p(X)$, $1 < p < \infty$; i.e. if (f_n) is an X-valued martingale with difference sequence (d_n) -which means that $d_1 = f_1$, and $d_n = f_n - f_{n-1}$ for $n > 1$ - then

$$\| \pm d_1 \pm d_2 \pm \ldots \pm d_n \|_{L_p(X)} \leq c_p \| d_1 + \ldots + d_n \|_{L_p(X)}.$$

This property, introduced by Burkholder [44], is independent of the number p, and is equivalent to the existence of a symmetric biconvex function ζ on $X \times X$ satisfying $\zeta(0, 0) > 0$ and $\zeta(x, y) \leq \|x+y\|$ if $\|x\| \leq 1 \leq \|y\|$. The U.M.D. property is a stronger form of super-reflexivity; it is inherited by subspaces, quotients and dual spaces, and it passes from X to $L_p(X)$ for $1 < p < \infty$.

Theorem 5.22.a. *The U.M.D. property is not a 3SP property.*

The proof is a result of Kalton outlined in [208]. It needs a fairly complete knowledge of interpolation theory, rearrangement invariant function spaces (Köthe function spaces); and blends several highly nontrivial results of Bourgain, Pisier and Lozanovskii. For this reason we shall only give an sketch.

The key for the proof is a result of Kalton [208] that roughly asserts that a certain twisted sum (known as the derived space) of an interpolation space X_θ has *the UMD* then there is a "band" around θ in such a way that for all ν in that band the spaces X_ν have *UMD*. Another tool is the identity $(X^*, X)_{1/2} = L_2$ of Lozanovskii [253], valid for Köthe function spaces. Two more results that are needed: Pisier's theorem [288] asserting that given a super-reflexive Köthe function space X and $0 < \theta < 1$, there is a Köthe function space Y such that $X = (L_2, Y)_\theta$; and

Bourgain construction [33], for any $1 < p < 2$, of a Köthe function space on [0, 1] that is p-convex and p^*-concave (i.e., superreflexive) but not *UMD*. With all this, we pass to the

Sketch of the proof of theorem 5.22.a following [208]. Starting with Bourgain's example of a Köthe function space X that is super-reflexive but not *UMD*, find, via Pisier's result, some $Y(\theta)$ such that $(L_2, Y(\theta))_\theta$ fails the *UMD* for $0 < \theta < 1$. By combining isomorphic copies of the spaces $Y(1/n)$ as bands in a single space it can be shown that there is a Köthe function space X so that $[X^*, X]_\theta$ is *UMD* only if $\theta = 1/2$ when Lozanovskii's identity $(X^*, X)_{1/2} = L_2$ holds. By the key result of Kalton above mentioned, the corresponding derived space $d_\Omega L_2$ has not *UMD*. $\square$

5.23   Mazur rotations problem

The well-known problem of the rotations of Mazur is to know if a *separable* Banach space where the group of isometries acts transitively on the unit sphere (i.e., such that given two points in the unit sphere there is an isometry sending one into the other) must be a Hilbert space. A space with this property shall be called *isotropic*. There exist non-separable isotropic spaces, such as $L_p(\mu)$ for a homogeneous non-σ-finite measure μ, as well as separable isotropic non-complete spaces (simple functions in $L_p(\mathbb{R})$). A combination of Lusky [254] almost-isotropic embedding (any normed space can be embedded isometrically into some Banach space whose isometries act almost-transitively over the unit sphere of the former space) plus some use of ultraproducts yield that any normed space can be isometrically embedded into some isotropic space.

The isometric $3SP$ problem for isotropic spaces has a negative answer. The following results are due to F. Cabello [46].

Theorem 5.23.a. *If V is an isotropic space not isomorphic to a Hilbert space then $V \times l_2$ endowed with any monotone norm is not isotropic.*

Proof. The space $V \times l_2$ is not a Hilbert space and thus some hyperplane cannot be complemented by a norm one projection. On the other hand, if π is a hyperplane of l_2 then $V \times \pi$ is a hyperplane of $V \times l_2$ complemented by a norm one projection. We only need the following nice result [46, Prop. 6.1]:

Proposition 5.23.b. *An isotropic space is isometric to a Hilbert space if and only if there is some hyperplane complemented by a norm one projection.* $\square$

   This implies that

Theorem 5.23.c. *Being isotropic is not stable by products.*

Thus, a space such as $L_p(\mu) \times l_2$ is not isotropic under the product norm. Nevertheless, $L_p(\mu)$ is isotropic for a homogeneous non-σ-finite measure μ and $1 < p \neq 2$, and $L_p(\mu) \times l_2$ is in fact isomorphic to $L_p(\mu)$; thus, $L_p(\mu) \times l_2$ is isotropic under some equivalent renorming. So, the interesting 3*SP* question is:

Question. *Is to be isotropic under some equivalent renorming a 3SP property?*

There also exist examples of isotropic spaces $L_1(\mu)$ (μ a homogeneous non-σ-finite measure) or $C(K)$ (suitable ultrapower of almost-isotropic renormings of $C(K')$ spaces); see [47].

Some related notions are defined by Cabello in his thesis [46]. Consider a group G of isometries of X acting transitively on the unit sphere S. Given $p \in S$, if we set $G_p = \{T \in G : Tp = p\}$ then there is a natural set-isomorphism

$$G/G_p \rightarrow S.$$

We say that X is *continuously isotropic* when this map is an homeomorphism. The space X is said to be *lipschitz isotropic* when the map is moreover lipschitzian, i.e., $\|x - y\| \geq c \inf\{\|Id - T\| : T \in G, Tx = y\}$. The isomorphic 3*SP* problem for lipschitz isotropic spaces has a negative answer:

Theorem 5.23.d. *To be lipschitz-isotropic under some equivalent renorming is not a 3SP property.*

Proof. The sequence $0 \rightarrow l_2 \rightarrow Z_2 \rightarrow l_2 \rightarrow 0$ gives a counterexample, where the nontrivial point is to prove that Z_2 s not lipschitz isotropic. This is consequence of the following result [46, Prop. 8.11]

Proposition 5.23.e. *The norm of a separable lipschitz isotropic space is C^∞-smooth.*

On the other hand (see [82, Thm. 4.1 p.209]) a C^∞-smooth Banach space not containing c_0 is of type 2. That Z_2 does not contain c_0 is a typical 3*SP* result, while that Z_2 is not of type 2 was proved in 5.1.a.   □

Question. *Is the space Z_2 isotropic?*

5.24   Never condensed spaces

Let us say that a Banach space X is *never condensed* if given any infinite dimensional subspace Z there is a decomposition $X = A \oplus B$ such that both $A \cap Z$ and

$B \cap Z$ are infinite dimensional. This property is studied in [75] under the name of property (B). Spaces with unconditional basis and subprojective spaces (i.e., spaces where every infinite dimensional subspace contains an infinite dimensional complemented subspace) are never condensed. The space $C[0, 1]$ is not never condensed. By extension, one may think that finite-dimensional spaces are never condensed.

Question. *Is never condensation a 3SP property?*

The property of never condensation is related to the problem of decomposing a space into subspaces. Dixmier [95] proved that if H is a Hilbert space and $T \in L(H)$ is an operator with dense range then there are operators $T_1, T_2 \in L(H)$ with dense range such that $Im\, T = Im\, T_1 + Im\, T_2$ and $Im\, T_1 \cap Im\, T_2 = \{0\}$. Trying to generalize this result Cross, Ostrovskii and Schevchik [75, Thm. 5.3] obtained

Proposition 5.24.a. *A space X is never condensed if and only if for every separable space Y and all operators $T \in L(X, Y)$ injective with nonclosed dense range there are dense nonclosed subspaces M_1, M_2 that are the ranges of some operators such that*

$$Im\, T = M_1 + M_2 \ \text{and}\ M_1 \cap M_2 = \{0\}.$$

Chapter 6

Homological properties

Following Grothendieck's algebraic approach, many properties of Banach spaces may be characterized in terms of the coincidence of two operator ideals or two classes of bounded sets. This chapter can be divided in two parts, one considering properties defined in terms of operator ideals and the other considering properties defined in terms of bounded sets. It is true that there is a certain correspondence between the two (see [10]); but in this regard, we have organized the sections following tradition.

In the first part, we consider contravariant properties defined by two operator ideals A and B, in the sense that for every Banach space Y, $A(X,Y) \subset B(X,Y)$ or, equivalently, $id_X \in BA^{-1}$. Typical examples of properties of this type are the Dunford-Pettis, the Schur or the Orlicz properties. In the second part we consider properties of Banach spaces defined in terms of classes of bounded sets that are preserved by continuous operators, such as the Gelfand-Phillips or the BD property.

6.1 C_p properties, $1 \leq p \leq \infty$

Following [55] and [64], an operator T is said to be *p-converging*, $1 \leq p \leq \infty$, if it transforms weakly p-summable sequences (∞-summable sequences are understood as weakly null sequences) into convergent sequences. These operator ideals were introduced in [64]. The ideal of p-converging operators is denoted C_p. Clearly,

$$Sp(C_\infty) = \{X: id_X \in C_\infty\} = \text{Schur spaces};$$
$$Sp(C_1) = \{X: id_X \in C_1\} = \text{spaces not containing } c_0.$$

The following result is clear.

Theorem 6.1.a. *For all* $1 \leq p \leq \infty$, *Space(C_p) is a 3SP ideal.*

Proof. Let $0 \rightarrow Y \rightarrow X \rightarrow Z \rightarrow 0$ be a short exact sequence. Let (x_n) be a weakly p-

summable sequence in X. The sequence (qx_n) is weakly p-summable in Z, and thus norm convergent to 0. Passing to a subsequence, if necessary, one can select elements $y_n \in Y$ with $\| x_n - y_n \| \le 2^{-n}$ so that (y_n) is a weakly p-summable sequence in Y and then also convergent to 0. Clearly, the sequence (x_n) must converge to 0.   □

This proof explicitly appears in [68]. A functorial approach or a proof via lifting are not difficult to do.

6.2   Orlicz properties

An operator is said to be (p, q)-*summing*, $1 \le q \le p < \infty$, if it transforms weakly q-summable sequences into absolutely p-summable sequences. The operator ideal of (p, q)-summing operators is denoted $\Pi_{p, q}$. An operator is said p-*summing* if it is (p, p)-summing. The operator ideal of p-summing operators is denoted Π_p.

A Banach space is said to have *Orlicz property* if, for some p, every weakly-1-summable sequence is absolutely p-summable, i.e. $id_X \in \Pi_{p,1}$. Observe that if a Banach space has Orlicz property then the quantities

$$r_n(X) = inf_{\|x_1\| = \ldots = \|x_n\| = 1} \; sup_{\pm} \; \| \textstyle\sum_{i=1}^{n} \pm x_i \|$$

increase to ∞ when $n \to \infty$. This yields that X cannot contain l_∞^n uniformly; i.e., the space X is C-convex. We already know that C-convexity (or, equivalently, having finite cotype) is a $3SP$ property because it is the super property corresponding to *not containing* c_0. That C-convexity is equivalent to Orlicz property is a very remarkable result of Talagrand [344]. Precisely, Talagrand proves:

Proposition 6.2.a. *For $q > 2$, X has cotype q if and only if $id_X \in \Pi_{q,1}$.*

which implies that Orlicz property is equivalent to having finite cotype. By the way, Talagrand also proved in the twin paper [343] that *cotype 2 $\ne$ $id_X \in \Pi_{2,1}$*.

Here we sketch a simpler proof that Orlicz property is a $3SP$ property due to Rakov [301]. Set

$$O(X) = inf \{p : id_X \in \Pi_{p,1}\}.$$

The proof starts with the estimate

$$O(X) = \lim_{n \to \infty} \frac{\log n}{\log r_n(X)}.$$

Proof. The existence of that limit is proved by Rakov in [300]. To show the inequality $\geq$, let (I_n) be a sequence sets of n consecutive naturals, $max\ I_n < min\ I_{n+1}$, and observe that if $\{x_i : i \in I_n\}$ is a set of n norm one elements of X for which the quantities r_n are "almost" reached, the elements

$$y_i = \frac{x_i}{r_n} \frac{1}{n^{1+\varepsilon}}, \qquad i \in I_n$$

satisfy $\qquad \| \sum y_i \| \leq \sum_n \| \sum_{I_n} y_i \| \leq \sum_n \frac{1}{n^{1+\varepsilon}} < +\infty.$

On the other hand, if $i \in I_n$, they have norm

$$\| y_i \| = r_n^{-1} n^{-(1+\varepsilon)}.$$

This implies that

$$\sum_i \| y_i \|^p = \sum_n r_n^{-p} n^{1-(1+\varepsilon)p}.$$

This series diverges if for some $\varepsilon_1 > 0$

$$r_n^{-p} n^{1-(1+\varepsilon)p} \geq \varepsilon_1 n^{-1}.$$

Therefore, if for large n

$$p \leq \frac{\log n}{\log r_n}(1-\delta)$$

then it is possible to find ε and ε_1 so that the last series diverges. It has been proved

$$O(X) \geq \lim_{n \to \infty} \frac{\log n}{\log r_n(X)}.$$

The other inequality can be obtained analogously. The second estimate one needs is

$$r_{m+n}(X) \geq \frac{r_n(Z)}{m + r_m(Y) + r_n(Z)}.$$

Proof. Let $\{x_i\}_{1 \le i \le n+m}$ be a set of $n+m$ elements, $\| x_i \| = 1$, and let $0 < \delta < 1$. If $q: X \to X/Y$ denotes, as usual, the quotient map, let $A = \{ i: \| qx_i \| < \delta \}$ and $B = \{ i: \| qx_i \| \ge \delta \}$. It is clear that either *card* $A \ge m$ or *card* $B \ge n$.

Case 1, card $A \ge m$. For $i \in A$, pick elements $y_i \in Y$ such that $\| x_i - y_i \| < \delta$. Observe that $\| y_i \| > 1 - \delta$. This gives

$$max \left\| \sum_{i=1}^{m+n} \epsilon_i x_i \right\| \ge max \left\| \sum_{i \in A} \epsilon_i x_i \right\|$$

$$\ge max \left(\left\| \sum_{i \in A} \epsilon_i y_i \right\| - \left\| \sum_{i \in A} \epsilon_i (x_i - y_i) \right\| \right)$$

$$= max \left\| \sum_{i \in A} \epsilon_i y_i \right\| - m\delta$$

$$\ge (1-\delta) r_n(Y) - m\delta .$$

Case 2, card $B \ge n$. In this case

$$max \left\| \sum_{i=1}^{m+n} \epsilon_i x_i \right\| \ge max \left\| \sum_{i \in B} \epsilon_i x_i \right\|$$

$$\ge max \left\| q \left(\sum_{i \in B} \epsilon_i x_i \right) \right\|$$

$$\ge \delta \, r_n(Z).$$

Thus, one has

$$r_{m+n}(X) \ge min \left\{ (1-\delta) r_m(Y) - m\delta , \, \delta r_n(Z) \right\}.$$

It only remains an exercise in real function: maximize the function on the right hand side of that inequality: since $f(\delta) = \delta b$ is a straight line with positive slope and $g(\delta) = (1-\delta)a - m\delta$ is a straight line with negative slope, the maximum value of the minimum occurs at their intersection

$$\delta = \frac{a}{m + b + a}$$

and is

$$\frac{ab}{m + b + a}.$$

With those ingredients, one can prove a nice typical three-space inequality

$$O(X) \le O(Y) \, O(X/Y),$$

that gives

Theorem 6.2.b. *The Orlicz property is a 3SP property.*

However, since *having cotype q* is not a 3*SP* property (5.1.b) and Talagrand's result [344] one has.

Theorem 6.2.c. *For fixed p,* $1 < p < \infty$, *the p-Orlicz property* $id_X \in \Pi_{p,1}$ *is not a 3SP property.*

6.3   Grothendieck's theorem

It is said that a Banach space X satisfies Grothendieck's theorem if all operators from X into a Hilbert space are absolutely 1-summing, i.e.,

$$L(X, H) = \Pi_1(X, H)$$

which comes from Grothendieck's theorem that $\mathcal{L}_1$ spaces have this property.

Theorem 6.3.a. *Satisfying Grothendieck's theorem is a 3SP property.*

Proof. Recall the ideas around 2.5.a. Let $0 \to Y \to X \to Z \to 0$ be a short exact sequence where it is assumed that Y and Z satisfy Grothendieck's theorem. Let H be a Hilbert space. Since the functor *Hom*(. , H) is left exact, the sequence

$$0 \to L(Z, H) \to L(X, H) \to L(Y, H)$$

is exact. Since Z and Y satisfy Grothendieck's theorem, the sequence

$$0 \to \Pi_1(Z, H) \to L(X, H) \to \Pi_1(Y, H)$$

is exact. But 2-summing operators factorize through a L_∞-space (a fortiori, 1-summing operators do). The map $Y \to L_\infty$ can be extended to a map $X \to L_\infty$. This means that the sequence

$$0 \to \Pi_1(Z, H) \to L(X, H) \to \Pi_1(Y, H) \to 0$$

is exact. But also, the extension $X \to H$ is again 1-summing since it factorizes through L_∞; therefore the sequence

$$0 \to \Pi_1(Z, H) \to \Pi_1(X, H) \to \Pi_1(Y, H) \to 0$$

is exact. The three-lemma proves that the middle map is surjective and thus X satisfies Grothendieck's theorem.   □

Jarchow's approaches this result by writing

$$\mathcal{F}_2^{-1} \circ \Pi_1 = \mathcal{F}_2^{-1} \circ [\Pi_1 \circ \mathcal{F}_\infty]^{inj}$$

and applying the method of 2.7 to conclude that it is a 3*SP* property. He then poses the question of characterizing those Banach spaces that satisfy the p-Grothendieck theorem, i.e.,

$$id_X \in \mathscr{F}_2^{-1} \circ \Pi_p = \mathscr{F}_2^{-1} \circ [\Pi_p \circ \mathscr{F}_\infty]^{inj} , \ 1 < p < \infty.$$

Applying method 2.7 one obtains that those are $3SP$ ideals.

6.4   Hilbert-Schmidt spaces

This property was introduced by Jarchow in [184]. A Banach space X is a *Hilbert-Schmidt space* if the operators $l_2 \to l_2$ that factorize through X are precisely the Hilbert-Schmidt operators. Jarchow denotes $\Pi_{2,2,2}$ the largest operator ideal whose components on Hilbert spaces are exactly the Hilbert-Schmidt operators. Since

$$\Pi_{2,2,2} = \mathscr{F}_2^{-1} \circ \Pi_2 = \mathscr{F}_2^{-1} \circ [\mathscr{F}_2 \circ \mathscr{F}_\infty]^{inj},$$

it follows from Jarchow's factorization method 2.7 that $Sp(\Pi_{2,2,2})$ is a $3SP$ ideal.

Theorem 6.4.a. *Hilbert-Schmidt spaces form a $3SP$ ideal.*

Although having cotype 2 is not a $3SP$ property, Jarchow mentioned to us that it follows from results of Stareuli [333] that *Hilbert-Schmidt spaces of cotype 2 form a $3SP$ ideal*. Observe that Hilbert-Schmidt spaces of cotype 2 are precisely those cotype 2 spaces verifying Grothendieck's theorem.

6.5   Radon-Nikodym properties

There are many equivalent formulations for the Radon-Nikodym property of a Banach space. The most classical one comes in terms of differentiability of X-valued functions. We choose one which haves a clear operator theoretic flavour: this allows an attack to the corresponding $3SP$ problem via lifting for operators (see 2.7.).

Recall that an operator $L_1[0,1] \to X$ is said to be *representable* if

$$Tf = \int_{[0,1]} f \, g \, dm$$

where m is the Lebesgue measure in $[0, 1]$ and $g \in L_\infty([0,1],X)$. A Banach space X is said to have the *Radon-Nikodym property* (*RNP*, in short) if every operator T: $L_1[0,1] \to X$ is representable.

The following characterization of representable operators is in [93, III.1.8]:

Proposition 6.5.a. *An operator $T: L_1 \to X$ is representable if and only if it factorizes through l_1.*

The following proof is in [106].

Theorem 6.5.b. *The Radon-Nikodym property is a 3SP property.*

Proof. Let $0 \to Y \to X \to Z \to 0$ be a short exact sequence, and assume that Y and Z have *RNP*. Let $T: L_1 \to X$ be a continuous operator. The composition qT factorizes through l_1 as $qT=BA$. The map $B: l_1 \to Z$ can be lifted to a map $C: l_1 \to X$ such that $qC=B$. The map $T-CA: L_1 \to Y$ can also be factorized through l_1, and this gives a factorization of T through l_1. $\square$

6.5.c. The Analytical *RNP*
was introduced by Bukhvalov and Danilevich [43] for a complex Banach space in terms of the differentiability of those X-valued measures μ on $\{z \in \mathbb{C}: |z|=1\}$ whose Fourier coefficients of negative index vanish. The *ARNP* is equivalent [100] to the property that for every operator $T: L_1/H_0^1 \to X$ the product TQ factorizes through l_1; here $Q: L_1 \to L_1/H_0^1$ is the quotient map and H_0^1 denotes the subspace of L_1 formed by those functions f such that $\mathscr{F}f(n)=0$ for all $n \le 0$, and $\mathscr{F}$ is the Fourier transform.

The following partial answer to the 3SP problem for the *ARNP* is in [99]:

Proposition 6.5.d. *If Y is a closed subspace of X such that Y has the RNP and X/Y has the ARNP then X has the ARNP.*

Proof (analogous to the proof for the *RNP*). Let $0 \to Y \to X \to Z \to 0$ be a short exact sequence where Y has *RNP* and Z has *ARNP*. Given $T: L_1/H_0^* \to X$ a continuous operator the composition qTQ factorizes through l_1 as $qTQ=BA$. The map $B: l_1 \to Z$ can be lifted to a map $C: l_1 \to X$ such that $qC=B$. The map $TQ-CA: L_1 \to Y$ can also be factorized through l_1, and this gives a factorization of TQ through l_1. $\square$

Question. Recall that $RNP^{dual} = $ Asplund. *What property is $ARNP^{dual}$?*

6.5.e. The Near *RNP*
was introduced in [219] as an attempt to understand the properties of the Volterra operator $V: L_1[0,1] \to C[0,1]$ given by

$$Vf(t) = \int_{[0,\,t]} f.$$

The Volterra operator is completely continuous but not representable. Nevertheless, Bourgain [30] proved that given any completely continuous operator $D: L_1 \to L_1$ the composition VD is representable.

Taking this as starting point, an operator $T: L_1 \to X$ is said to be *nearly representable* if for every completely continuous operator $D: L_1 \to L_1$ the composition TD is representable. A Banach space X is said to have the *near RNP* if every nearly representable operator $L_1 \to X$ is representable.

The same method of proof of 6.5.b and 6.5.d provides:

Theorem 6.5.f. *The Near RNP is a 3SP property.*

Proof. Let $0 \to Y \to X \to Z \to 0$ Y be a short exact sequence where Y and Z have the NRNP. Given a nearly representable operator $T: L_1 \to X$ the composition qT is representable and factorizes through l_1 as $qT = BA$. The map $B: l_1 \to Z$ can be lifted to a map $C: l_1 \to X$ such that $qC = B$. The map $T - CA: L_1 \to Y$ can also be factorized through l_1, and this gives a factorization of T through l_1.   □

6.5.g The weak RNP was introduced by Musial [263]. It is an open question if it is a 3SP property. In [263] and [264] it is proved that $(\text{weak-RNP})^{dual}$ is equivalent to not containing l_1. Thus

Theorem 6.5.h. $(\text{weak-RNP})^{dual}$ *is a 3SP property.*

6.6   Dunford-Pettis properties

A Banach space X is said to have the *Dunford-Pettis property* (DPP) if every weakly compact operator $T: X \to Y$ transforms weakly compact sets of X into relatively compact sets of Y. Equivalently, if given weakly null sequences (x_n) and (f_n) in X and X^* respectively, $\lim f_n(x_n) = 0$. Clearly, if X^* has the DPP then X has the DPP. It is an open problem to know when the DPP passes from X to X^*. It is easily deduced from Rosenthal's lemma that if X does not contain l_1 then X has the DPP if and only if X^* is Schur. Also, there is the example $l_1(l_2^n)$ of Stegall that has the DPP (it is a Schur space) while its dual $l_\infty(l_2^n)$ has not since it contains l_2 complemented, and complemented subspaces of DPP spaces are DPP spaces. Typical examples of spaces with the DPP are $C(K)$ spaces [158] and, thus, L_1-spaces. Since $\mathcal{L}_\infty$ spaces (resp. $\mathcal{L}_1$-spaces) are those whose bidual is complemented in some $C(K)$ space (resp. L_1-space), then $\mathcal{L}_\infty$-spaces and $\mathcal{L}_1$-spaces have DPP.

For some time it has been believed that the DPP was something like the opposite of reflexivity, until Bourgain and Delbaen [36] constructed a somewhat reflexive $\mathcal{L}_\infty$-space (and thus having *the DPP*). It was also believed that the absence of reflexive subspaces or quotients could be sufficient conditions, until Bourgain and Delbaen [36] showed an Orlicz sequence space without reflexive subspaces and lacking the Dunford-Pettis property, and Leung [236] exhibited an ad hoc space (see 6.6.m) without the Dunford-Pettis property and without reflexive quotients. Besides, there exist Orlicz sequence spaces without the Dunford-Pettis property and without reflexive quotients (see [36]). Schreier space (see 4.6.a) could also work since it is c_0-saturated as well as all its quotients (proved by Odell [268]) and does not have the Dunford-Pettis property (see below).

Theorem 6.6.a. *The Dunford-Pettis property is not a 3SP property.*

The proof shall be given in 6.6.f. since the same counterexample works for the hereditary Dunford-Pettis property.

Problem. *Is DPP ^{dual} a 3SP property ?*

Positive results regarding Dunford-Pettis properties are:

Proposition 6.6.b. *If Y does not contain copies of l_1, and Y and X/Y have the DPP then X has the DPP.*

Proof. Let (x_n) be a weakly null sequence in X and let (f_n) be a weakly null sequence in X^*. The sequence $(f_n | Y)$ is weakly null in Y^*, which is a Schur space; thus it is norm null. Since $Y^* = X^*/Y^\perp$, there is a weakly null sequence (g_n) in $Y^\perp$ with $\lim \|f_n - g_n\| = 0$. Since $Y^\perp = (X/Y)^*$, from where $\lim g_n(x_n) = 0$, it easily follows

$$\lim f_n(x_n) = \lim g_n(x_n) + \lim (f_n - g_n)(x_n) = 0. \qquad \square$$

This result is in Stehle's thesis [332]. Another proof, only valid for the sequence $0 \to X \to X^{**} \to X^{**}/X \to 0$, was given in [62].

Proposition 6.6.c. *If Y has the DPP and X/Y is Schur then X has the DPP.*

Proof. Again, let (x_n) be a weakly null sequence in X and (f_n) a weakly null sequence in X^*. Since X/Y is Schur, the sequence $(x_n + Y)$ is norm null. If (y_n) is a sequence in Y such that $\lim \|x_n - y_n\| = 0$, it is weakly null. Therefore $\lim f_n(y_n) = 0$ and $\lim f_n(x_n) = \lim f_n(y_n) + \lim (x_n - y_n) = 0. \qquad \square$

The hereditary Dunford-Pettis property. A Banach space X is said
to have the *hereditary* Dunford-Pettis property (DPP_h) if every closed subspace of X has the *DPP*.

Proposition 6.6.d. *A Banach space X has the DPP_h if every weakly null sequence (x_n) admits a weakly 1-summable subsequence.*

Proof. Firstly observe that if no normalized weakly null sequences exist in X this is a Schur space and then it obviously has the DPP_h. Therefore the word "normalized" can be added to the statement without risk.

Necessity. Let (x_n) be a normalized weakly null sequence in X, which can be considered a basic sequence. Consider a weak*-null sequence (f_n) in $[(x_n)]^*$ such that $f_i(x_j) = \delta_{ij}$. No subsequence of (f_n) can be weakly Cauchy since, otherwise, $[(x_n)]$ would fail the *DPP*; thus, some subsequence (f_m) of (f_n) is equivalent to an l_1-basis. This makes the operator T: $[x_m] \to c_0$ defined by $T(x)=(f_m(x))$ an isomorphism.

Sufficiency. If (x_n) is a weakly 1-summable sequence with constant K and (f_n) is a bounded sequence of functionals then $\lim f_n(x_n)=0$:

$$sup_k \sum_{n=1}^{k} |f_N(x_n)| = sup_k \sum_n f_N(\pm x_n)$$

$$= sup_k f_N\left(\sum_n \pm x_n\right)$$

$$\leq \|f_N\| \; sup_k \left\|\sum_{n=1}^{k} \pm x_n\right\|$$

$$\leq \|f_N\| \, K .$$

$\square$

This gives a well known result of Elton [109]

Corollary 6.6.e. *An infinite dimensional Banach space contains either c_0, l_1 or a subspace without the DPP property.*

Proof. Since the space is either Schur or contains a normalized weakly null basic sequence. $\square$

and also the following equivalent formulation for the DPP_h (see also [89]):

$$DPP_h = Sp(\text{ weak } W_1),$$

where *weak W_1* denotes the ideal of operators transforming weakly null sequences into sequences having weakly-1-summable subsequences (see section 4.6; also [67]).

The following result is in [60].

Theorem 6.6.f. *The Dunford-Pettis and the Hereditary Dunford-Pettis are not 3SP properties.*

Proof. In 4.6.a it was already proved that *weak-W_1* is not a 3SP property. This settles the situation for the hereditary Dunford-Pettis property. Recall that the counterexample was the diagonal short exact sequence induced by the pull-back of the arrows: the canonical inclusion $i: S \to c_0$ and a quotient map $q: l_1 \to c_0$; that is

$$0 \to Ker(q+i) \to l_1 \oplus S \to c_0 \to 0,$$

where S is the Schreier's space [325].

One only has to verify that S has not even the Dunford-Pettis property. Recall that S is the space obtained by completion of the space of finite sequences with respect to the norm:

$$\|x\|_S = \sup_{[A\ admissible]} \sum_{j \in A} |x_j|,$$

where a finite subset of natural numbers $A = \{n_1 < \ldots < n_k\}$ is said to be *admissible* if $k \leq n_1$. Also, recall that S is algebraically contained in c_0, that it has an unconditional basis formed by the canonical vectors (e_i), and that a sequence (x^n) of S is weakly null if and only if, for every j, the sequence $(x_j^n)_{n \in \mathbb{N}}$ is null.

In this way, that S does not possess the *DPP* is consequence from the fact that the dual unit vector sequence is weakly null in S^*: and this immediately follows from the estimate (see [60])

$$\| \sum_{k=1}^{2^N} e_{i_k} \|_{S^*} \leq \sup_{\|x\| \leq 1} \sum_{k=1}^{2^N} |x_{i_k}|$$

$$\leq \sup_{\|x\| \leq 1} |x_{i_1}| + |x_{i_2} + x_{i_3}| + |x_{i_4} + \ldots + x_{i_7}| + \ldots$$

$$\ldots + |x_{i_{2^{N-1}}} + \ldots + x_{i_{2^N}}|$$

$$\leq N.$$

Another form to prove that S has not the *DPP* is showing that S^{**} is also contained in c_0: this shows that the inclusion $S \to c_0$ is weakly compact; and it is certainly not completely continuous.   □

The role of S can be played by any Banach space F having a weakly null basis and the following properties: *i)* F is continuously contained into c_0, *ii)* F has not the Dunford-Pettis property, and *iii)* Any normalized weakly null sequence of blocks with $\|\ \|_\infty$ norms decreasing to 0 contains a weakly 1-summable subsequence. Leung's space [236] (see also 6.6.m) or the *lunatic space* [54] are different examples.

Proposition 6.6.g. $(DPP_h)^{dual}$ *is a 3SP property.*

Proof. It turns out that $X \in (DPP_h)^{dual}$ if and only if X^* has the Schur property: since if X^* has the DPP_h and does not have Schur property then, by 6.6.d, it must contain c_0 and thus l_∞; in which case X^* contains subspaces without the Dunford-Pettis property, and this is a contradiction. $\qquad\square$

6.6.h. Qiyuan [299] considers *Schauder basis determining properties* as those properties such that X has P if every subspace with basis has P, and asks (open problem 2, p.97) if Schauder basis determining properties are $3SP$ properties. The answer is no: the DPP_h is determined by subspaces with basis.

The "hereditary by quotients" Dunford-Pettis property This

property: every quotient of X has the DPP is, in some sense, dual of the DPP_h.

Theorem 6.6.i. *The "hereditary by quotients" Dunford-Pettis property (DPP_{hq}, in short) is a 3SP property.*

Proof. A Banach space with this property cannot contain l_1 (since a space having l_1 as a subspace has l_2 as a quotient). Since X^* has the Schur property if and only if X has the DPP and does not contain l_1, a Banach space X has the DPP_{hq} if and only if X^* has the Schur property, which is a $3SP$ property. $\qquad\square$

Observe that it has been proved that $DPP_{hq} = (DPP_h)^{dual}$.

The Dunford-Pettis property of order p, $1 < p \le \infty$. The

Dunford-Pettis property can be described by means of the inclusion $W(X,Y) \subset C_\infty(X,Y)$. A weaker form is obtained replacing C_1 by C_p: Let $1 \le p \le \infty$. We shall say as in [64] that a Banach space X has the *Dunford-Pettis property of order p* (in short DPP_p) if $W(X,Y) \subset C_p(X,Y)$ holds for any Banach space Y. Obviously, DPP_p implies DPP_q when $q < p$. Also, $DPP = DPP_\infty$ and every Banach space has DPP_1. It follows from the definition that if $id(X) \in C_p$ then X has the DPP_p, and that if $id(X) \in W_p$ then X does not have the DPP_p. The following result is in [64].

Proposition 6.6.j. *For a given Banach space X, they are equivalent: X has DPP_p ($1 \le p \le \infty$) and If (x_n) is a weakly-p-summable sequence of X and (x_n^*) is weakly null in X^* then $(x_n^* x_n) \to 0$.*

Some Banach spaces having those properties, with optimal values of p are: $C(K)$ and L_1 have the DPP_∞. If $1 < r < \infty$, l_r has the DPP_p for $p < r^*$. If $1 < r < \infty$, $L_r(\mu)$ has the DPP_p for $p < \min(2, r^*)$. Tsirelson's space T has the DPP_p for all $p < \infty$. However, since T is reflexive, it does not have the DPP. Tsirelson's dual space T^* does not have the DPP_p for any $p > 1$.

It is open to know if the Dunford-Pettis property of order p is a $3SP$ property.

Reciprocal Dunford-Pettis property. It is apparently unknown whether the *reciprocal DPP* (completely continuous operators are weakly compact; i.e., for all spaces Y, $C_\infty(X, Y) \subset W(X, Y)$; also shortened sometimes to *RDPP*) is a $3SP$ property.

Somewhat DP property. *Somewhat DP* spaces are those such that every subspace contains a subspace with the Dunford-Pettis property. Stehle proves in [332] that *somewhat DP* spaces form a $3SP$ ideal.

Theorem 6.6.k. *Somewhat DP is a 3SP property.*

Proof. If $\Lambda = \{$*spaces without infinite dimensional DP subspaces*$\}$ then

$$\Lambda(i) = \text{ somewhat DP},$$

and apply the incomparability method.   □

If, instead, one starts with

$$DP = \{\text{spaces with the Dunford-Pettis property}\}$$

then since all $C(K)$ spaces have the Dunford-Pettis property and every Banach space is isomorphic to a subspace of some $C(K)$,

$$DP(i) = \{\text{finite dimensional spaces}\}$$

and

$$DP(ii) = \{\text{all Banach spaces}\};$$

and if

$$DP_H = \{\text{spaces such that every subspace has the Dunford-Pettis property}\}$$

then

$$DP_H(i) = \{\text{spaces containing no infinite dimensional subspace with } DPP_h\},$$

that, after the result of Elton 6.6.e, leads us to

$$DP_H(i) = \{c_0\}(i) \cap \{l_1\}(i) = \{c_0, l_1\}(i)$$

and thus

$$DP_H(ii) = \{c_0, l_1\}\text{-saturated spaces.}$$

This class is contained in the class of *somewhat DPP* spaces, but we do not know whether they coincide. In fact, if there is some space

$$X \in somewhat\ DPP \setminus \{c_0,\ l_1\}\text{-saturated}$$

then X contains no unconditional basic sequence, because if (x_n) is a basic unconditional sequence in X then $[x_n]$ does not contains c_0 or l_1 and must be reflexive and cannot contain an infinite dimensional DP space. It may be therefore of some interest to ask

Problem. *Does there exist some Banach space such that every subspace contains a subspace with DP and another without DP?*

and maybe to suggest that constructions of Gowers-Maurey type [156,157] are candidates to be examinated.

Surjective Dunford-Pettis property. It was introduced by Leung in

[238] as follows: every surjective weakly compact operator is completely continuous. It is a weaker form of the DPP. The following result is a combination of results in [238] and [27].

Proposition 6.6.1. *For a given Banach space X, they are equivalent:*
  i) X has the surjective DPP, and
  ii) If (x_n) is a weakly null sequence of X and (f_n) is weakly null in X* spanning a reflexive subspace, then $\lim (f_n\ x_n) = 0$.

Theorem 6.6.m. *The surjective Dunford-Pettis property is not 3SP.*

Proof. Let us consider *Leung's space* [238] as follows: A finite subset $A \subset \mathbb{N}$ is said to be Leung-admissible if $card\ A \leq min\ A$ and $card(A) = 2^i$ for some $i \geq 0$. The space L is defined as the completion of finite sequences (x_n) under the norm

$$\| (x_n) \| \ = \ \sup\ \| A(x_n) \|\ _{\sqrt{i}}$$

where $A(x_n)$ is the sequence equal to (x_n) if $n \in A$ and equal to zero otherwise, $\|\ \|_p$ is the l_p-norm and the supremum is taken over all Leung-admissible sets. The space L has the surjective Dunford-Pettis property but not the Dunford-Pettis property [236]. We shall follow [27] to show that $l_1 \oplus L$ has not the $sDPP$. Otherwise, let $T: L \to R$ be an operator taking values on a separable reflexive space, and let $q: l_1 \to R$ be a quotient map. The map $T+q$ should be completely continuous, as well as T. In conclusion, L should have the DPP, arriving to a contradiction.   □

Once it is known that Schreier's space does not have reflexive quotients or DPP

it could also work as a counterexample: $S \oplus l_1$ does not have the *sDPP*.

6.6.n. Leung also defined in [239] the *weak Dunford-Pettis property* in Banach lattices as follows: every weakly compact operator sends disjointly supported weakly null sequences into norm null sequences. It is unknown to us if this is a $3SP$ property in the domain of Banach lattices (where $3SP$ should have a slightly different meaning).

Dieudonné property.

A Banach space X has the *Dieudonné property* (*D*) if *Dieudonné operators* (i.e., operators transforming weakly Cauchy sequences into weakly convergent sequences) are weakly compact. It is unknown to us whether this is a $3SP$ property.

Moreover, it is unknown whether the reciprocal Dunford-Pettis property (*RDPP*) and the Dieudonné property (*D*) are equivalent. Clearly, $D \Rightarrow RDPP$. A space with *RDPP* without *D* should contain l_1 (otherwise all weakly completely continuous operators on X would be weakly compact, and the space would have *D*) but not complemented (a projection would be completely continuous not weakly compact, and the space would fail *RDPP*).

Grothendieck property and its reciprocal.

A Banach space X is said to have the *Grothendieck property* if weak* null sequences in X^* are weakly null. Equivalently, operators having separable range are weakly compact. Wheeler [357] defined the *reciprocal Grothendieck property* as follows: every weakly compact operator has separable range and gave it the name of *Dunford-Pettis-Phillips property* (shortened by Wheeler to DP^3).

Theorem 6.6.o. *The Grothendieck property and DP^3 are $3SP$ properties.*

Proof. That the Grothendieck property is a $3SP$ property has already been obtained through a lifting result in 2.4.e. As for the DP^3, let Y be a subspace of X such that Y and X/Y have DP^3. If $T: X \to M$ is weakly compact, then $\overline{TY}$ is separable. The induced operator $T_q: X/Y \to M/\overline{TY}$ is weakly compact, and thus $\overline{TX}/\overline{TY}$ is separable. Since separability is $3SP$ property, $\overline{TX}$ must be separable.   □

The proof could also follow noting that X has the Grothendieck property if and only if $S_1(X) = X^{**}$ and applying the 3-lemma.

Other related properties.

A property related to the Dunford-Pettis property was pointed out in [184]. We shall say that a Banach space X has property *WAC* if, for every Banach space Y, every weakly compact operator $X \to Y$ is *absolutely continuous* (*AC*) in the sense of Niculescu [266]. After a characterization of Jarchow and Pelczynski [182, Thm. 20.7.6 and Corol. 20.7.7], an operator is absolutely continuous if and only if it belongs to the injective hull of the closure of the ideal of p-summing operators, $p \geq 1$ (it does not matter which p is chosen). It is unknown whether property WAC is a $3SP$ property. The counterexample $l_1 \times S$ which worked for the DPP could also work for property WAC, were it true that $Ker(q+i)^{**}$ has DPP. It is clear that $WAC \Rightarrow DPP$. Moreover, it is an elementary observation that if X has DPP and does not contain l_1 then it has WAC. In general it is false that $DPPh \Rightarrow WAC$: it was proved by Niculescu that $L(l_1, X) \subset AC(l_1 X)$ implies X super-reflexive. Taking as X a reflexive non-superreflexive Banach space, one sees that l_1 has not property WAC.

6.7   Covariant properties

Let us briefly consider covariant properties, in the sense that they are defined as

$$\text{For all Banach spaces } Y, \, A(Y, X) \subset B(Y, X).$$

Here A and B are two operator ideals defined by transforming sequences of a given type (the same for both operator ideals) into sequences of, respectively, types a and b. Regarding $3SP$ problems, this type of properties behave in the same form as the corresponding property $a \subset b$ and thus things depend on the existence of a lifting result involving sequences of types a and b.

For instance, a simple although not completely trivial result.

Theorem 6.7.a. *The property that for all spaces Y all Banach-Saks operators $X \rightarrow Y$ are compact is a 3SP property.*

Proof. There exist lifting for convergent sequences. □

K_p **properties**. Jarchow defines in [185] that a Banach space X has *property K_p*, $1 \leq p \leq \infty$, if $L(L_p, X) = K(L_p, X)$ (i.e., all operators from L_p into X are compact). The following results are in [185]

Proposition 6.7.b. $K_1 = $ *Schur property. For $1 < q < p < 2 \leq r < \infty$ one has the following containments*: $K_1 \subset K_q \subset K_p \subset K_2 = K_r$, *and also* $K_1 \subset K_\infty \subset K_2$. *All inclusions are proper except* $K_\infty \subset K_2$, *which is an open problem.*

This type of properties were also considered in [55], obtaining:

Proposition 6.7.c. *X has K_p if and only if $id_X \in C_{max\{2, p^*\}}$.*

Therefore, by 6.1,

Theorem 6.7.d. K_p *is a 3SP property.*

As an application of this result, one obtains:

Proposition 6.7.e. *There are reflexive as well as non-reflexive Banach spaces belonging to all classes K_p, $1 < p$ and not to K_1.*

Proof. For the reflexive case use Tsirelson space: every Banach-Saks operator is compact. For K_∞ use Diestel-Seifert result that all operators from $C(K)$ into T are Banach-Saks [92].

The non-reflexive case uses the inductive limit method 1.8.b to obtain a short exact sequence: $0 \to T \to \mathcal{L}_\infty(T) \to \mathcal{L}_\infty(T)/T \to 0$, where T is Tsirelson's space. The quotient space has *RNP* and Schur properties. The property *Banach-Saks operators are compact* is a 3SP property (6.7.a) and thus the middle space has that property. It does not have Schur since it contains T. $\square$

Question (of [185]). *Does there exist a space with the Banach-Saks property belonging to K_p for all $1 < p \leq \infty$?*

6.8   The Gelfand-Phillips property

A Banach space X is said to have the *Gelfand-Phillips property* if limited sets are relatively compact. A subset $A \subseteq X$ is said to be limited if, whenever (f_n) is a weak* null sequence of X^*, $\lim_{n \to \infty} \sup \{ |f_n(x)| : x \in A \} = 0$. Following [321] the property was originated by an error: the following statement of Gelfand [126] and Mazur: "A subset K of a Banach space X is relatively compact if and only if each weakly* null sequence of X^* converges uniformly on K." It is not difficult to see that only the only if part is true. It was pointed out by Phillips [284] that the if part is false: the unit basis of c_0 in l_∞ is, definitely, not relatively compact although $\lim_{n \to \infty} |f_n(e_n)| = 0$ for each weakly*-null sequence (f_n) of $(l_\infty)^*$ (since weakly and weakly* null sequences in $(l_\infty)^*$ coincide (Grothendieck property), and l_∞ has the Dunford-Pettis property.

The following result was proved by Bourgain and Diestel [37].

Proposition 6.8.a. *Limited sets in Banach spaces are conditionally weakly compact; i.e., every sequence in a limited set admits a weakly Cauchy subsequence.*

Proof. Assume A is a bounded set in a Banach space X which is not conditionally weakly compact. Then A contains a sequence (x_n) having no weakly Cauchy subsequences. By Rosenthal theorem, we may assume that (x_n) is equivalent to the unit vector basis of l_1. So it generates a subspace isomorphic to l_1. Now since the inclusion map from l_1 into c_0 factorizes through l_2, it is absolutely summing [248, Thm. 266]. So is admits an extension $T:X \to c_0$. Now, if (f_n) is the image under T^* of the unit vector basis of l_1, we have that (f_n) is a weakly null sequence in X^* and $f_n(x_n)=1$ for all n. Hence A is not limited.   □

The following result is useful to check if a Banach space has the Gelfand-Phillips property.

Proposition 6.8.b. *A Banach space X has the Gelfand-Phillips property if and only if every limited weakly null sequence in X is norm null.*

Proof. Clearly, Banach spaces having the Gelfand-Phillips property satisfy the condition. Conversely, let us assume that X fails the Gelfand-Phillips property. So it contains a non-relatively compact limited subset A. Now there exists $\varepsilon > 0$ and a sequence (x_n) in A so that $\|x_m-x_n\| > \varepsilon$, for $m \neq n$. Since the set A is weakly conditionally compact, we may assume the sequence (x_n) is weakly Cauchy. So taking $y_n = x_{n+1}-x_n$, we obtain a limited weakly null sequence (y_n) in X which does not converges in norm to 0.   □

The next two results provide examples of spaces with the Gelfand-Phillips property.

Proposition 6.8.c. *Spaces having weak*-sequentially compact dual unit ball have the Gelfand-Phillips property.*

Proof. Assume that X is a Banach space with weak*-sequentially compact dual ball, and let (x_n) be a limited weakly null sequence in X. Let us consider the operator T: $l_1 \to X$ given by $Te_n=x_n$, where (e_n) stands for the unit vector basis of l_1. Since $T(Ball\ l_1)$ is limited, the dual operator $T^*: X^* \to l_\infty$ is weak*-to-norm sequentially continuous, and so $T^*(B_{X^*})$ is compact. Thus, the operator T is compact and (x_n) is norm null.

Proposition 6.8.d. *Let K be a sequentially compact, compact space. Then $C(K)$ has the Gelfand-Phillips property.*

Proof. As in the previous proof, take a limited weakly null sequences (f_n) in $C(K)$, consider the operator $T: l_1 \to C(K)$ given by $Te_n=f_n$, and observe that $T(Ball(l_1))$ is a limited subset of $C(K)$. Since weak* convergent sequences in a dual space are uniformly convergent on limited sets, the dual operator T^* sends weak* convergent sequences into norm convergent sequences. Thus, T^* transforms $\{\delta_k: k \in K\}$ into a relatively compact subset of X. Hence, T^* is compact and so is T.   □

In order to show that the Gelfand-Phillips property is not a $3SP$ property, recall the construction of the Stone compact (see 3.5) generated by an algebra of subsets of N. The example will be a refinement of the counterexample given in 4.8.a to the $3SP$ problem for having weakly* sequentially compact dual ball. The results of this section are taken from Schlumprecht [321] and [323].

So, our aim is to describe a Stone compact over an algebra Ξ so that $C(K_{\Xi})/c_0$ has the Gelfand-Phillips property, but $(1_{\{n\}})$ is a limited set in $C(K_{\Xi})$; in particular $C(K_{\Xi})$ fails the Gelfand-Phillips property.

Let ω_1 denote the first uncountable ordinal and let $P_\infty(N)$ denote the class of all infinite subsets of N. Recall that two sets $M, N \in P_\infty$ are said to be almost disjoint if $M \cap N$ is finite; we say that M is almost contained in N if $M \backslash N$ is finite.

For a well-ordered family $(R_\alpha)_{\alpha < \Omega}$ in $P_\infty(N)$, where Ω is an ordinal, let us consider the following properties:

(F)   If $\alpha < \beta$ then either $R_\beta \backslash R_\alpha$ is finite or $R_\alpha \cap R_\beta$ is finite.

(FM)   The family $(R_\alpha)_{\alpha < \Omega}$ is a maximal family satisfying (F); i.e., it satisfies (F) and for every $R_\Omega \in P_\infty(N)$ the family $(R_\alpha)_{\alpha \le \Omega}$ does not satisfy (F).

Lemma 6.8.e. *Let $(R_\alpha)_{\alpha < \Omega} \subset P_\infty(N)$ be a family satisfying (FM). Then the cardinality of the family equals the cardinality of the continuum.*

Proof. First we show that for $\alpha < \Omega$ there are $\alpha_1, \alpha_2 \in (\alpha, \Omega)$ so that R_{α_1} and R_{α_2} are almost disjoint and almost contained in R_α. In fact, given a set $N \in P_\infty(R_\alpha)$ with $R_\alpha \backslash N$ infinite, by (FM) we find $\beta < \Omega$ so that $N \cap R_\beta$ and $N \backslash R_\beta$ are both infinite. Clearly β has three properties: $\alpha < \beta$, $R_\beta \backslash R_\alpha$ is finite and $R_\alpha \cap R_\beta$ is infinite.

Define α_1 to be the minimum of all β having these three properties. We repeat the process for $N = R_\alpha \backslash R_{\alpha_1}$ and get α_2 so that $(R_\alpha \backslash R_{\alpha_1}) \backslash R_{\alpha_2}$ and $(R_\alpha \backslash R_{\alpha_1}) \cap R_{\alpha_2}$ are both infinite, and again α_2 satisfies the above three properties. By the minimality of α_1 we have $\alpha_1 < \alpha_2$, and $R_{\alpha_2} \backslash R_{\alpha_1}$ infinite implies $R_{\alpha_2} \cap R_{\alpha_1}$ finite. In this way the assertion stated at the beginning of the proof is proved.

Now we construct inductively a tree

$$\{\alpha(n, j) : n \in N \cup \{0\} \ ; j = 1, ..., 2^n\}$$

so that for every $R_{\alpha(n, j)}$, its successors $R_{\alpha(n+1, 2n)}$ and $R_{\alpha(n+1, 2n+1)}$ are, respectively, almost disjoint and almost contained in $R_{\alpha(n, j)}$. For every branch γ in the tree,

$$\gamma = \{\alpha(n, j(\gamma, n)) : n \in N \cup \{0\} \ ; j(\gamma, n+1) \in \{2j(\gamma, n), 2j(\gamma, n) - 1\}\}$$

and for every $n \in N \cup \{0\}$ we choose an integer

$$K(\gamma, n) \in \bigcap_{m \leq n} R_{\alpha(m, j(\gamma, m))}$$

(note that this intersection is infinite), so that $N_\gamma = \{K(\gamma, n) : n \in N\}$ is infinite. By the choice of the sets $R_{\alpha(n, j(\gamma, n))}$ it follows that for two different branches γ and γ' the intersection $N_\gamma \cap N_{\gamma'}$ is finite. By the maximality, for each branch γ one has $\alpha(\gamma) < \Omega$ so that $N_\gamma \backslash R_{\alpha(\gamma)}$ and $N_\gamma \cap R_{\alpha(\gamma)}$ are both infinite. So it follows from (F) that for each branch γ and each $n \in N$ the set $R_{\alpha(\gamma)} \backslash R_{\alpha(n, j(\gamma, n))}$ is finite.

All this implies that the sets $R_{\alpha(\gamma)}$ are pairwise almost disjoint, in particular different. Since the set of all branches has the cardinality of the continuum, the proof is finished. $\qquad\square$

We say that a family $(R_\alpha)_{\alpha < \omega_1} \subset P_\infty(\mathbb{N})$ satisfies (C) if for each $N \in P_\infty(\mathbb{N})$ there is an $\alpha < \omega_1$ with $R_\alpha \subset N$. Observe that if a well-ordered family satisfies (F) and (C) then it also satisfies (FM).

Proposition 6.8.f. *There exists a family* $(R_\alpha)_{\alpha < \omega_1} \subset P_\infty(\mathbb{N})$ *satisfying properties (F) and C.*

Proof. Let $(N_\alpha)_{\alpha < \omega_1}$ be a well-ordering of $P_\infty(\mathbb{N})$. By transfinite induction, if for every $\alpha < \omega_1$ one chooses an infinite subset R_α of N_α so that (F) holds for $(R_\beta)_{\beta \leq \alpha}$, we are done.

Assume R_α has already been chosen for each $\alpha < \beta$, $\alpha < \omega_1$, and put

$$I := \{\beta < \alpha : R_\beta \cap N_\alpha \text{ is infinite}\}$$

The set I is well-ordered by the order on $[0, \alpha]$ and the family $(R_\beta \cap N_\alpha)_{\beta \in I}$ satisfies (F) as a subset of $P_\infty(N_\alpha)$. Since I is countable, using 6.8.e it is possible to find an infinite subset $R_\alpha \subset N_\alpha$ so that $(R_\alpha \cap N_\alpha)_{\beta \in I \cup \{\alpha\}}$ still satisfies (F). Clearly $(R_\beta)_{\beta < \alpha+1}$ satisfies (F), and the proof is complete. $\qquad\square$

Proposition 6.8.g. *Let K be the Stone compact associated to a family $(R_\alpha)_{\alpha < \omega_1} \subset P_\infty(\mathbb{N})$ satisfying (F) and (C).*

(a) *The unit ball of $(C(K)/c_0)^*$ is weak*-sequentially compact.*

(b) *The sequence $(1_{\{n\}})$ forms a limited set.*

Proof. a) Since $C(K)/c_0$ is isomorphic to $C(K\backslash N)$, by 6.8.d it is enough to show that

$$K\backslash N = \{\delta_x : x \in K\backslash N\} \subset C(K\backslash N)^*$$

is sequentially compact.

Let us denote $S_\alpha = $ *support* $(1_{R_\alpha}) \subset K$. The sets $(S_\alpha \backslash N)_{\alpha < \omega_1}$ form a subbasis of the topology of $K\backslash\mathbb{N}$. By (F), it follows that for $\alpha < \beta < \omega_1$ one has that either

$$(S_\alpha\backslash\mathbb{N}) \cap (S_\beta\backslash\mathbb{N}) = \varnothing \quad or \quad S_\alpha\backslash\mathbb{N} \subset S_\beta\backslash\mathbb{N}..$$

So, every point $p \in K\backslash\mathbb{N}$ has $\{S_\alpha\backslash\mathbb{N} : p \in S_\alpha\}$ as a neighborhood basis, well-ordered by inclusion. Hence, each separable closed subset of $K\backslash\mathbb{N}$ is metrizable (given a sequence of points (p_n), p_n is excluded from neighborhoods of p_1 for $\alpha > \alpha_n$; since no countable subset of ω_1 is cofinal, p_1 has a countable fundamental system of neighborhoods: those with index $\alpha \leq max \, \alpha_n$; repeat the argument for the other p_n to obtain that every point has a countable fundamental system of neighborhoods) and the space $K\backslash\mathbb{N}$ is sequentially compact.

   b) Let us denote $f_n = \mathbf{1}_{\{n\}}$, and assume (μ_n) is a sequence in the unit ball of $C(K)^* = M(K)$ with $r = \lim \, sup_{n\to\infty} \, \mu_n(f_n) > 0$. It is enough to prove that the sequence (μ_n) is not weak*-convergent to 0.

   By Rosenthal's Lemma (see [89, p.82]) there is $M \in P_\infty(\mathbb{N})$ such that the subsequence $(f_n)_{n \in M}$ satisfies

$$\mu_n(f_n) = \mu_n(\{n\}) > \frac{3}{4}r,$$

and

$$|\mu_n|(M\backslash\{n\}) < \frac{1}{4}r,$$

for every $n \in M$. By property C of the family (R_α) it is possible to select $\alpha < \omega_1$ such that $R_\alpha \subset M$. Now, taking $1_{R_\alpha} \in C(K)$ one has that for every $n \in R_\alpha$

$$|\mu_n(1_{R_\alpha})| = |\mu_n(R_\alpha)| \geq |\mu_n\{n\}| - |\mu_n|(M\backslash\{n\}) > \frac{r}{2},$$

and the sequence (μ_n) is not weak*-null. $\square$

   A consequence of this result and 6.8.d is

Theorem 6.8.h. *The Gelfand-Phillips property is not a 3SP property.*

6.9   Properties (u), (V) and (V*)

The Orlicz-Pettis theorem implies that every weakly compact operator is unconditionally converging; i.e., it takes weakly 1-summable, also called weakly unconditionally Cauchy (in short *wuC*), sequences into unconditionally converging series. In [108], Pelczynski considered the spaces X such that the converse implication is true and called them spaces with property (V). In that paper he also introduced the properties (u), and (V^*) as follows: Let X be a Banach space. It is said that X has property (*u*) if, for every weakly Cauchy sequence (x_n) in X, there exists a *wuC* sequence (z_n) in X such that the sequence $(x_n - \sum_{j=1}^{n} z_j)$ is weakly null. It is said that X has property (*V*) if, for every Banach space Z, every unconditionally converging operator from X into Z is weakly compact; equivalently, whenever K is a bounded subset of X^* such that

$$lim_{n\to\infty} \, sup\{|f(x_n)| : f \in K\} = 0$$

for every *wuC* series $\sum_n x_n$ in X, then K is relatively weakly compact. A Banach space X is said to have property (*V**) if whenever K is a bounded subset of X such that

$$lim_{n\to\infty} \, sup\{|f_n(x)| : x \in K\} = 0$$

for every *wuC* series $\sum_n' f_n$ in X^*, then K is relatively weakly compact.

Pelczynski [108] proved that properties (*u*) and (*V**) are preserved by taking subspaces and property (*V*) is preserved by taking quotients. Moreover, that spaces $C(K)$ have property (*V*) and the spaces $L_1(\mu)$ have property (*V**). Clearly, c_0 has property (*u*), but an space which is not weakly sequentially complete and contains no copies of c_0 (such as James' space J) fails property (*u*). Since J is isomorphic to a subspace of $C[0,1]$ this space also fails property (*u*).

Proposition 6.9.a. *If a Banach space X has property (u) and contains no copies of l_1 then it has property (V).*

Proof. Let Y be a Banach space and let $T:X \to Y$ be an unconditionally converging operator. Given a bounded sequence (x_n) in X, by the hypothesis there exist a subsequence $(x_{n(k)})_{k \in \mathbf{N}}$ and a *wuC* series $\sum y_i$ in X such that the sequence

$$(x_{n(k)} - \Sigma^k y_i)$$

is weakly null. So $(Tx_{n(k)})$ is weakly convergent to ΣTy_i and the operator T is weakly compact.   □

The following result is in [61].

Theorem 6.9.b. *Properties (u) and (V) are not 3SP properties.*

Proof. A counterexample is provided by the spaces X_p, $1 \le p < \infty$, of Figiel, Ghoussoub and Johnson [127] which we describe now. Denoting by $\mathbf{N}^* = \mathbf{N} \cup \{\infty\}$, and by c the space of converging sequences, set $X = l_1(c)$, i.e, the space of doubly-indexed sequences $a = (a_{ij})$, $i \in \mathbf{N}$, $j \in \mathbf{N}^*$, such that

$$lim_{j\to\infty} a_{ij} = a_{i\omega} \quad \text{and} \quad \|a\|_X = \sum_{i=1}^{\infty} (sup_j |a_{ij}|) < \infty .$$

Let $f^n \in X$ be given by $f^n_{ij} = 1$, if $i \le n \le j$, and 0 otherwise. The gauge $\| \ \|$ of the closed absolutely convex solid hull of the union of $Ball(X)$ and the sequence (f^n) is a lattice norm on X. For $1 \le p < +\infty$,

$$\|x\|_p = \| \, |x|^p \, \|^{1/p}$$

defines a new lattice norm on X, whose completion shall be denoted X_p.

Lemma 6.9.c. *For $1 < p < \infty$ the space X_p contains no copies of l_1*

Proof. Observe that from the definition of $\| \, . \, \|_p$ it follows that for any disjoint bounded sequence (y^n) in X_p one has

$$\left\| \sum_n a_n \, y^n \right\|_p \le C \, \| (a_n) \|_{l_p}$$

and thus (y^n) cannot be equivalent to the unit vector basis of l_1.

Let S_m be the projection on X_p defined by

$$(S_m x)_{ij} = x_{ij} \quad \text{for } i \le m;$$
$$(S_m x)_{ij} = 0, \text{ otherwise.}$$

It is clear that $S_m(X_p)$ is isomorphic to c_0. So given a sequence (x^n) in X_p equivalent to the unit vector basis of l_1, it is possible to find a sequence (z^n) of normalized blocks of (x^n) such that

$$lim_{n \to \infty} \, S_m \, z^n = 0$$

for all m, and it is possible to find a disjoint sequence (y^n) so that

$$lim \, \| z^n - y^m \| = 0,$$

in contradiction with the observation at the beginning of the proof. $\square$

Let $T: X \to c_0$ be the operator defined by $T(a_{ij}) = (a_{i\omega})$, and let T_p be its continuous extension to X_p. The proof of the theorem finishes with the following

Lemma 6.9.d. *For $1 < p < \infty$ the operator T_p is unconditionally converging and surjective. Moreover, $Ker \, T_p$ has property (u).*

Proof. Suppose that T_p is not unconditionally converging. Then it is an isomorphism on a subspace E of X_p isomorphic to c_0. Take a normalized basis (z^n) for E that is k-equivalent to the unit vector basis of c_0. Since X is dense in X_p there is no loss of generality assuming that each z^n lies in X. One has

$$\| T_p \, z^n \| = max_i |z^n_{i\omega}|, \quad \text{and} \quad lim_{n \to \infty} \, z^n_{i\omega} = 0 \quad \text{for each } i \in N.$$

Hence, it is possible to find a $\delta > 0$ and $i_1 < i_2 < \dots$ so that $|z^n_{i_n \omega}| > \delta$ for all n.

Now, if we define $y_{ij}^n = |x_{ij}|$ for $i = i_n$ and $y_{ij}^n = 0$ otherwise, then the sequence (y^n) in X_p is equivalent to the unit vector basis of c_0; and taking $x_{ij}^n = (y_{ij}^n)^p$ we obtain a sequence (x^n) in X_1 equivalent to the unit vector basis of c_0.

Next, for each n, select j_n so that $x_{i_n j_n} > \delta^p$. Passing to a subsequence, if necessary, assume that $i_{n+1} > j_n$ for each n. Observe that if (f^n) is the sequence of vectors appearing in the definition of the norm then $f_{i_n j_n}^n = 0$ for all n. So,

$$\| \textstyle\sum_{n=1}^m x^n \|_1 \geq \delta^p m,$$

arriving to a contradiction.

In order to show that $Ker(T_p)$ has property (u), let (x^n) be a weakly Cauchy sequence in $Ker(T_p)$. Since the limit

$$x_{ij} = lim_{n \to \infty} x_{ij}^n$$

exists for every (i,j), its is possible to define

$$y_{ij}^k = \begin{cases} x_{ij}, & \text{for } k=i \, ; \\ 0, & \text{otherwise}. \end{cases}$$

Clearly $y^k = (y_{ik}^k)$ belongs to $Ker(T_p)$, and $\Sigma_k y^k$ is a wuC series in $Ker(T_p)$ such that

$$(x^n - \textstyle\sum_{k=1}^n y^k)$$

is weakly null. Hence $Ker(T_p)$ has property (u). $\qquad\square$

In [134], it is shown that properties (V) (resp. (V^*)) verify a restricted version of the $3SP$ property, namely, when X/Y (resp. Y) is reflexive.

Problem. *Is (V^*) a $3SP$ property?* A counterexample, if it exists, should be a weakly sequentially complete Banach space, not isomorphic to a subspace of a Banach lattice, and whose dual is not weakly sequentially complete.

6.10 Property (w)

Saab and Saab introduced in [318] *property (w)* as: a Banach space X has property (w) if every continuous operator $T: X \to X^*$ is weakly compact. In [318], it is proved that every $\mathcal{L}_\infty$-space and every space with property (V) has property (w). Moreover, property (w) is equivalent to property (V^*) in Banach lattices. Nevertheless,

Theorem 6.10.a. *Property (w) is not a 3SP.*

Proof. In [238], Leung constructed a non-reflexive Banach space (not a Banach lattice) E with property (w) such that E^* also has property (w). Since if E is not reflexive then the natural map

$$E \times E^* \to E^{**} \times E^*$$

is not weakly compact, and thus $E \times E^*$ can never have property (w). □

6.11 Other properties

Property (BD). After the paper [36], Emmanuele isolated the property that bounded sets of X where weak* null sequences converge uniformly are weakly compact and called it *property BD*. It is unknown to us whether *BD* is a 3SP property.

Property L. Considered by Emmanuele in [111]. A bounded subset K of X^* is a *L-set* if weakly null sequences of X converge uniformly on K. *Property L* is defined as: *L*-sets are relatively weakly compact.

Question. *Is L a 3SP property ?*

It was proved by Bator in [19, Thm. 2] that *L*-sets are relatively compact if and only if the space does not contain l_1. □

Dunford-Pettis sets are compact. This property was considerd by Emmanuele [113]. Teresa Alvarez suggested to us:

Question. *Is the property "Dunford-Pettis sets are relatively compact" a 3SP property?*

We remark that its dual *is* a 3SP property since it has been proved by Bator [19, Corol. 7] that it is equivalent to not containing l_1.

Chapter 7

Approximation Properties

A classical result in functional analysis (see [248]) asserts that if the Banach space X has a Schauder basis then every compact operator $T \in K(Y, X)$ is limit in norm of finite rank operators. Spaces verifying this property are said to have the Approximation Property (AP). The question whether any Banach space X has the AP was open for a long time, and it was finally solved in the negative by Enflo [115], providing also a negative solution to the basis problem: has every separable Banach space a basis? Several variants of the AP were introduced during the effort to solve these problems; and after Enflo's example of a Banach space failing the AP, the study of these properties by its own gained added importance. The most fruitful of those has been the Bounded Approximation Property (BAP). It has been known since 1971 (see [248, Thm. 1.e.13]) that a Banach space has the BAP if and only if it is isomorphic to a complemented subspace of a Banach space with basis. In [336] has found a space with the BAP without a basis.

Here we consider a couple of properties related to the AP, and show that none of these are 3SP properties. In fact, the counterexamples are very strong: *Every separable Banach space X contains a subspace Y such that both Y and X/Y have a Finite Dimensional Decomposition* [195]; and *if X contains a copy of c_0 then it is possible to select Y in such a way that both Y and X/Y have a basis* [255]. Moreover, a result due to Szarek [335] implies that having a basis is not stable under taking complemented subspaces: there exist reflexive Banach spaces X, Y such that $X \times Y$ and Y have a basis, but X has not basis.

We shall also show counterexamples for some unconditional versions of the approximation properties, and finally we present some positive results for restricted versions of the 3SP property for AP, and BAP due to Godefroy and Saphar [135].

A Banach space X is said to have the *Approximation Property* if, for every compact subset K of X and $\varepsilon > 0$, there exists a finite-rank operator $T: X \to X$ such that $\|x - Tx\| < \varepsilon$ for all $x \in K$. The space X has the *Bounded Approximation*

Property if the finite-rank operators in the definition of the *AP* may be taken uniformly bounded; the Banach-Steinhaus theorem yields the equivalent formulation: there exists a sequence (B_n) of finite rank operators in X pointwise convergent to the identity; i.e., $lim_{n \to \infty} B_n x = x$, for all $x \in X$. The space X is a π-*space* if there exists $C > 0$ such that every finite dimensional subspace of X is contained in a C-complemented finite dimensional subspace; equivalently, if the operators B_n in the definition of the *BAP* may be taken to be projections. The space X is said to admit a *Finite Dimensional Decomposition* (*FDD*) if the finite-rank operators (B_n) in the definition of the *BAP* may be selected verifying $B_n B_m = B_{min\ (m,n)}$. The space X has a (Schauder) *basis* if it has a *FDD* such that the sequence (B_n) consists of projections with n-dimensional range. One has

$$basis \Rightarrow FDD \Rightarrow \pi\text{-}space \Rightarrow BAP \Rightarrow AP \ .$$

All the converse implications, but one, are known to be false: Enflo obtained an example of a separable Banach space without AP, and Davies [77] showed that such subspaces exist inside c_0 or l_p $1 < p < \infty$, $p \neq 2$. Figiel and Johnson [121] proved that the AP does not imply the BAP. Szarek [336] gave an example of a Banach space with a *FDD* and without basis. Read in [304] showed that BAP does not imply π-space. It is still an open question whether π-space implies *FDD*.

Throughout this chapter, $L(X)$ denotes the space of all linear continuous operators $X \to X$; while F(X), resp. K(X), denote the subspace of finite rank (resp. compact) operators. Also, if A is a subset of a Banach space X then [A] denotes the closure of the linear span of A. If (x_j) is a sequence of points then the closure of the linear span of $\{x_1, \dots x_n\}$ shall often be written as $[x_j]_{j=1}^n$.

7.1   Spaces with Finite Dimensional Decompositions

The following result shall be useful for the construction of the counterexamples.

Proposition 7.1.a. *Let X be a separable Banach space. There exist sequences (x_n) in X and (f_n) in X^* such that $f_i(x_j) = \delta_{ij}$, $[x_n] = X$ and $\|x\| = \sup\ \{|f(x)| : f \in [f_i] \}$.*

Proof. Take a sequence (y_k) dense in the unit sphere of X and another sequence (g_k) in the unit sphere of X^* so that $g_n(y_n) = 1$. It is enough to construct biorthogonal sequences (x_n) and (f_n); i.e., such that $f_i(x_j) = \delta_{ij}$, in such a way that

$$[x_i]_{i=1}^{2n} \supset [y_j]_{j=1}^n \quad \text{and} \quad [f_i]_{i=1}^{2n} \supset [g_i]_{i=1}^n .$$

In order to obtain those sequences, put $x_1 = y_1$ and $f_1 = g_1$. Then, assuming that x_i and f_i have been selected for $i < 2n$, take the smallest integer p such that $g_p \notin$ span $\{f_1, .., f_{2n-1}\}$, and let

$$f_{2n} = g_p - \sum_{i=1}^{2n-1} g_p(x_i) f_i .$$

Moreover, selecting k such that $f_{2n}(y_k) \neq 0$, let

$$x_{2n} = f_{2n}(y_k)^{-1}\left(y_k - \sum_{i=1}^{2n-1} f_i(y_k) x_i\right).$$

Let q be the smallest integer such that $y_q \notin \mathrm{span}\{x_1, .., x_{2n}\}$, and let

$$x_{2n+1} = y_q - \sum_{i=1}^{2n} f_i(y_q) x_i .$$

Moreover, selecting k such that $g_k(x_{2n+1}) \neq 0$, put

$$f_{2n+1} = g_k(x_{2n+1})^{-1}\left(g_k - \sum_{i=1}^{2n} g_k(x_i) f_i\right).$$

It is immediate to verify that the sequences (x_n) and (f_n) satisfy the appropriate conditions, and thus the proof is complete. $\qquad\square$

The next result provides examples showing that the *AP*, the *BAP*, to be a π-space or to have a *FDD* are not *3SP* properties.

Proposition 7.1.b. *Let X be a separable Banach space. Then there exists a subspace Y of X such that both Y and X/Y have a FDD.*

Proof. Take the sequences (x_n) in X and (f_n) in X^* provided by Proposition 7.1.a Then, choose inductively finite sets of integers $\sigma_1 \subset \sigma_2 \subset \ldots$ and $\tau_1 \subset \tau_2 \subset \ldots$ so that

$$\sigma = \bigcup_i \sigma_i \quad \text{and} \quad \tau = \bigcup_i \tau_i$$

are complementary infinite subsets of the positive integers, and for each positive integer n, and vectors $x \in [x_i]_{i \in \sigma_n}$ and $f \in [f_i]_{i \in \tau_n}$ one has

   i) $\|x\| \leq 2 \sup \left\{ |g(x)| : \|g\| = 1, \ g \in [f_i \ ]_{i \in \sigma_n \cup \tau_n} \right\}$

and

   ii) $\|f\| \leq 2 \sup \left\{ |f(y)| : \|y\| = 1, \ y \in [x_i]_{i \in \sigma_n \cup \tau_n} \right\}.$

We set

$$Y = [x_i]_{i \in \sigma} = {}^{\perp}[f_i]_{i \in \tau}.$$

Consider, for every n, the projections S_n and T_n in X defined by

$$S_n x = \sum_{i \in \sigma_n} f_i(x) x_i \quad \text{and} \quad T_n x = \sum_{i \in \tau_n} f_i(x) x_i .$$

Claim: For every n one has

$$\| S_n|_Y \| \le 2, \quad T_n|_Y = 0,$$

and for all $x \in X$, dist $(T_n x, Y) \le 2$ dist (x, Y).

Proof of the claim. For the first part, note that by (1) given $y \in Y$ it is possible to select

$$g \in [f_i]_{i \in \sigma_n \cup \tau_n}$$

so that $\| g \| \le 2$ and $\| S_n y \| = g(S_n y) = g(y)$. It is also clear that $T_n|_Y = 0$, and thus the remaining of the claim is equivalent to $\| T_n^*|_{Y^{\perp}} \| \le 2$. In order to show this, observe that if $g \in Y^{\perp}$, since $T_n^* \in [f_i]_{i \in \tau_n}$ by (2), it is possible to choose

$$y \in [x_i]_{i \in \sigma_n \cup \tau_n}$$

so that $\| y \| \le 2$ and $\| T_n^* g \| = (T_n^* g)(y) = g(y)$, and the claim is proved. □

Finally, observe that the claim implies that (S_n) induces a *FDD* in Y and (T_n) a *FDD* in X/Y.

Theorem 7.1.c. *AP, BAP, π-space or FDD are not 3SP properties.*

Proof. Apply the last theorem to any separable Banach space X failing the *AP*. □

Proceeding in an analogous way one can show that a Banach space X with X^* separable contains a subspace Y so that both $Y^{\perp} = (X/Y)^*$ and $X^*/Y^{\perp}$ have a *FDD*. We refer to [195] or to [248, 1.f.3] for details. From this it follows

Theorem 7.1.d. *The AP^{dual}, the BAP^{dual}, to have a dual π-space or to have a dual with FDD are not 3SP properties.*

Proof. Apply the result to a separable reflexive subspace failing the AP. □

7.2   Spaces with a basis

In this section we shall show that having a basis is not a *3SP* property. This result is due to Lusky [255]. Let (G_n) be a sequence of finite dimensional Banach spaces such that

(†) The sequence (G_n) is dense in F with respect to the Banach-Mazur distance; and

(‡) For each integer n there exists an infinite subset of integers J such that G_n is isometric to G_j for each $j \in J$.

We denote

$$C_p = l_p(G_n), \text{ if } 1 \le p < \infty, \text{ and } C_\infty = c_0(G_n),$$

endowed with the natural norms $\| . \|_p$ and $\| . \|_\infty$, respectively. Because of (‡), the space C_p is isometric to $l_p(C_p)$. Moreover, if (H_n) is another sequence of finite dimensional spaces satisfying (†) and (‡) then

$$d(C_p, l_p(H_n)) = 1.$$

So, the space C_p is essentially unique.

The following lemma is in [329, Corol. 9.2, p.289].

Lemma 7.2.a. *Given a finite dimensional space E there is a finite dimensional space F such that $E \oplus_\infty F$ has a basis with basis constant lesser than or equal to 5.*

Proposition 7.2.b. *Let $1 \le p \le \infty$. If X is a Banach space with a FDD, then $X \times C_p$ has a basis.*

Proof. Let (E_n) be a FDD for $X \oplus C_p$. By Lemma 7.2.a, for every n there exist finite dimensional spaces F_n such that $E_n \oplus_\infty F_n$ has a basis with basis constant ≤ 5. Let us consider the sequence (G_n) in the definition of C_p. For every n there exists i_n so that $d(F_n, G_{i_n}) \le 2$. Hence

$$E_n \oplus_\infty G_{i_n}$$

has a basis with basis constant ≤ 10, and

$$(E_n \oplus_\infty G_{i_n})$$

is a FDD for

$$(X \oplus C_p) \oplus_\infty l_p(G_{i_n}),$$

while the spaces $E_n \oplus G_{i_n}$ have bases with uniformly bounded basis constants. Hence the space

$$(X \oplus C_p) \oplus_\infty l_p(G_{i_n})$$

has a basis. Moreover

$$(X \oplus C_p) \oplus_\infty l_p(G_{i_n}) \simeq X \oplus l_p(G_n) \oplus l_p(G_{i_n}) \simeq X \oplus l_p(G_n) \simeq X \oplus C_p.$$

Whence $X \oplus C_p$ has a basis. $\square$

Since any finite dimensional Banach space is almost isometric to a subspace of some l_∞^k space, the spaces G_n in the definition of C_p can be considered as subspaces of suitable $l_\infty^{k_n}$ spaces. In this way, C_∞ is itself isometric to a subspace of

$$c_0 = c_0(l_\infty^{k_n}),$$

and $(C_\infty)^*$ is isometric to a quotient of l_1. Moreover, $(C_\infty)^*$ is isomorphic (even almost isometric) to C_1, and one has

$$c_0 / C_\infty \simeq c_0\left(l_\infty^{k_n}/G_n\right) \simeq c_0(H_n),$$

and thus

$$C_\infty^\perp = \left(c_0 / C_\infty\right)^* \simeq l_1(H_n^*).$$

Hence both spaces c_0/C_∞ and $C_\infty^\perp$ have a FDD.

Theorem 7.2.c. *To have a basis is not a 3SP property.*

Proof. Let X be a separable Banach spaces failing the AP. By Theorem 7.1.b there exists a subspace Y of X such that both Y and X/Y have a FDD. Since X does not have AP, the space $X \oplus c_0 \oplus l_1$ does not have a basis. However, since $Y \oplus C_\infty^\perp$ has a FDD, by 7.2.b

$$Y \oplus C_\infty \oplus C_\infty^\perp$$

has a basis; also

$$X/Y \oplus c_0/C_\infty \oplus l_1/C_\infty^\perp \simeq X/Y \oplus c_0/C_\infty \oplus C_1$$

has a basis, because $X/Y \oplus c_0/C_\infty$ has a FDD. $\square$

With a bit more of effort, a stronger result could be proved (see [255]).

Theorem 7.2.d. *Any separable Banach space containing a copy of c_0 has a subspace Y such that both spaces Y and X/Y have a basis.*

Question. *Is it possible to extend the previous theorem to any Banach space? In particular, what happens in the case of a separable dual Banach space?*

The following result of [48] is a consequence of the counterexample for "being a dual space" shown in 3.7.

Theorem 7.2.e. *To have a boundedly complete basis is not a 3SP property.*

Proof. Observe that spaces with boundedly complete basis coincide with dual spaces with basis. An examination of the construction in 3.7.b shows that the subspace and the quotient space have basis.   □

Theorem 7.2.f. *To have an unconditional basis is not a 3SP property.*

Proof. Recall that the space Z_2 has a subspace Y isomorphic to l_2 such that Z_2/l_2 is again isomorphic to l_2. However, it is proved in [213] that Z_2 does not contain complemented subspaces with unconditional basis.   □

A stronger result is available. Recall that a Banach spaces has *Gordon Lewis local unconditional structure* (G.L.l.u.st.if there exists a constant C such that for any finite dimensional subspace E of X the inclusion map $i: E \to X$ factors through some finite dimensional space U with maps $A: U \to X$ and $B: E \to U$, $i=AB$, in such a way that $\|A\| \|B\| un(U) \le C$, where $un(U)$ is the unconditional constant of U. It can be found in [194] that

Proposition 7.2.g. *The space Z_2 does not have G.L. l.u.st.*

And thus

Theorem 7.2.h. *To have G.L.l.u.st. is not a 3SP property.*

7.3   Positive results

Let X be a Banach space. We denote $L(X)$ the space of all linear bounded operators and $F(X)$ the space of all elements of $L(X)$ having finite dimensional range. Since the space $L(X^*)$ is isomorphic to the dual of the projective tensor product $(X^* \hat{\otimes}_\pi X)^*$ (see, e.g., [93]) we can consider in $L(X^*)$ the associated weak* topology. We denote by τ_k the topology on $L(X)$ of uniform convergence over compact sets of X.

It follows readily from the definition that a Banach space X has the *AP* if and only if id_X belongs to the τ_k-closure of F(X), and X has the *BAP* if and only of there exists $\lambda > 0$ such that id_X belongs to the τ_k-closure of $\lambda Ball(F(X))$. One has

$$(L(X), \tau_k)^* = X^* \hat{\otimes}_\pi X,$$

which means that any $z \in (L(X), \tau_k)^*$ has the form

$$z(T) = \sum_{i=1}^{\infty} f_i(Tx_i)$$

for some $(f_i) \subset X$ and $(x_i) \subset X^*$ with $\sum \|f_i\| \|x_i\| < \infty$; i.e., $z = \sum_i f_i \otimes x_i$ in the projective tensor product space (see [248, 1.1.3]).

Lemma 7.3.a. Let (R_α) be a net in $F(X^*)$ such that

$$weak^*\text{-}lim \, R_\alpha = id_{X^*}$$

There is a net (S_β) in $F(X)$ such that

$$weak^* -lim \, S_\beta^* = id_{X^*}$$

Moreover, if (R_α) is bounded then (S_α) can be selected bounded.

Proof. Every R_α has the form

$$R_\alpha(f) = \sum_{i=1}^{n} F_i(f) f_i$$

with $F_i \in X^{**}$ and $f_i \in X^*$. Since X is weak* dense in X^{**}, R_α belongs to the weak* closure of $\{S^*: S \in \mathcal{F}(X)\}$, and the result is clear.

In the case of bounded nets, one has

$$weak^*\text{-}lim \, R_\alpha = id_{X^*}$$

if and only if for every $x \in X$ and $f \in X^*$ one has $lim \, R_\alpha f(x) = f(x)$. If $\alpha = (A, B, \varepsilon)$, where A and B are finite sets of X and X^* respectively, $0 < \varepsilon < 1$, and $R_\alpha \in F(X^*)$ verifies

$$|R_\alpha f(x) - f(x)| < \varepsilon$$

for every $x \in A$ and $f \in B$, then the principle of local reflexivity allows us to select $S_\alpha \in F(X)$ such that $f(S_\alpha x) = R_\alpha(f(x))$ for every $x \in A$ and $f \in B$, with $\|S_\alpha\| \leq (1+\varepsilon)\|R_\alpha\|$. Clearly (S_α) is a bounded net and

$$weak^* \, lim \, S_\alpha^* = id_{X^*} \qquad \qquad \square$$

Lemma 7.3.b. Let X be a Banach space. Then

    i) X has the AP if and only if id_{X^*} belongs to the weak*-closure of $F(X^*)$.
    ii) X has the BAP if and only if there exists $\lambda > 0$ such that id_{X^*} belongs to the weak*-closure of $\lambda Ball(F(X^*))$.

Proof of i) Assume that X has the AP. If $(R_\alpha)_\alpha$ is a net of finite dimensional operators on X that is τ_k-convergent to id_X and let $z = \sum_i f_i \otimes x_i \in X^* \hat{\otimes}_\pi X$. Then, one has

$$lim_\alpha \langle R_\alpha^*, z \rangle = lim_\alpha \sum_{i=1}^{\infty} f_i(R_\alpha x_i) = \sum_{i=1}^{\infty} f_i(x_i)$$

hence

$$weak^* - lim\ R_\alpha^* = id_{X^*}.$$

Conversely, assume there exists $(S_\alpha) \subset F(X^*)$ weak*-convergent to id_{X^*}. By 7.3.a we can assume that $S_\alpha = R_\alpha^*$ for some $R_\alpha \in F(X)$. Since (R_α) is weakly convergent to id_X in $(L(X), \tau_k)$, id_X belongs to the τ_k-closure of F(X).

The proof of *ii*) is analogous using bounded nets. $\qquad\square$

Lemma 7.3.c. *Let X be a Banach space and Y a subspace of X such that X/Y has AP. If there exists a bounded net $(T_\alpha) \subset F(X)$ such that for each $x \in Y$ and $f \in X^*$*

$$lim_\alpha\ \langle f, T_\alpha x \rangle = \langle f, x \rangle,$$

then X has the AP.

Proof. Let T be a weak*-cluster point of $\{T_\alpha^*\}$ in $L(X^*)$. Obviously $<T^*f, x> = <f, x>$ for each $x \in Y$ and each $f \in X^*$. Now, if $j: Y^\perp \to X^*$ is the canonical inclusion, there exists an operator $D: X^* \to Y^\perp$ such that $T - id_{X^*} = jD$. Since X/Y has the *AP* it is possible to take a net (S_β) in F(X/Y) such that

$$weak^* - lim\ S_\beta^* = id_{Y^\perp}$$

in $L(Y^\perp)$. If $V_\beta = j\ S_\beta^* D$ then

$$weak^* - lim\ V_\beta = j\ D$$

in $L(X^*)$, which shows that $id_{X^*} = T - jD$ belongs to the weak*-closure of the set $\{T_\alpha^* - V_\beta\}$; hence, by 7.3.b, X has the *AP*. $\qquad\square$

Lemma 7.3.d. *Let X be a Banach space, and Y a subspace of X such that X/Y has the BAP. Then X has the BAP if and only if there exists a bounded net (T_α) in F(X) such that for each $x \in Y$ and $f \in X^*$*

$$lim_\alpha\ \langle f, T_\alpha x \rangle = \langle f, x \rangle.$$

Proof. The direct implication is clear. For the converse, repeat the proof of 7.3.c using bounded nets, and apply 7.3.b. $\qquad\square$

Theorem 7.3.e. *Let X be a Banach space, and Y a subspace of X such that Y has the BAP and $Y^\perp$ is complemented in X*. If X/Y has the AP (resp. BAP) then X has the AP (resp. BAP).*

Proof. Let (R_α) be a bounded net in F(Y) such that $lim_\alpha R_\alpha x = x$ for each $x \in Y$. If $i: Y \to X$ is the canonical inclusion, since $Y^\perp$ is complemented in X^* there exists an operator $P: Y^* \to X^*$ such that $i^*P = id_{Y^*}$. Now, if

$$R_\alpha = \sum_{i=1}^n g_i \otimes x_i \; ; \quad g_i \in Y^*, \; x_i \in Y$$

then define $S_\alpha \in F(X)$ by

$$S_\alpha = \sum_{i=1}^n P(g_i) \otimes x_i.$$

For every $x \in Y$ and $f \in X^*$ one has

$$lim_\alpha \langle f, S_\alpha x \rangle = lim_\alpha \langle f, R_\alpha x \rangle = \langle f, x \rangle.$$

Moreover, $\|S_\alpha\| \le \|P\| \|R_\alpha\|$, and the net (S_α) is bounded. So, it is enough to apply 7.3.c and 7.3.b to conclude the proof.   □

Corollary 7.3.f. *Let X be a Banach space, and let (P) be one of the properties:*

a) to have a basis,
b) to have a FDD,
c) to be a π-space,
d) to have the BAP.

If X and X^{**}/X have (P) then X^{**} and X^* have (P).

7.4   Skipped-blocking decompositions

In the study of certain geometrical properties related with the Radon-Nikodym property, Bourgain and Rosenthal [39] introduced the notion of Skipped-blocking decomposition. Several properties of separable Banach spaces can be characterized in terms of these decompositions.

Let us recall that given a sequence (G_i) of subspaces of a Banach space X we denote $[G_i]_{i=1}^\infty$ the closed subspace of X generated by the G_i, and for $1 \le m \le n \le \infty$ we denote $G[m,n]$ the subspace generated by $\{G_i : m \le i \le n\}$. With all this, let P be a class of *FDD*'s. A sequence (G_i) of finite dimensional subspaces of X is a *P-skipped blocking decomposition (P-SBD)* of X if

i) $X=[G_i]_{i=1}^\infty$

ii) $G_i \cap [G_j]_{j \ne i} =0$ for all i.

iii) If (m_k) and (n_k) are sequences of integers, with $m_k < n_k < m_{k+1}$ for every k, then $(G[m_k,n_k])$ is a *FDD* of class P for the subspace $[G \ [m_k,n_k]]_{k=1}^\infty$ of X.

We observe that a *P-SBD* need not to be a *FDD*. In fact, if we take as P the class of all *FDD*'s, then any separable Banach space has a *P-SBD* (see [32]).

Clearly, if X has a *SBD* of some kind, then X is separable. The following results are in [315, 1.2; 1.3; 4.1] and [39].

Proposition 7.4.a. *Let X be a separable Banach space.*
 i) X has Polish ball if and only if it has a reflexive-SBD;
 ii) X has the PCP if and only if X has a boundedly complete-SBD.
iii) X^* separable if and only if X has a shrinking-SBD.

And, therefore,

Theorem 7.4.b. *To have a reflexive (resp. boundedly complete, shrinking) SBD is a 3SP property.*

Proof. The corresponding properties (Polish ball, *PCP*, dual separable) are 3*SP* properties. □

Summary

The first column lists properties in a shortened form. The second says "yes" or "no" to the corresponding 3SP problem; ? means that it is an open problem. The third column contains additional information (equivalent formulations; a counterexample for the 3SP problem, if such exists; a method that works, or an equivalent property for which some method works, when the 3SP problem has a positive answer). A blank cell means that, although there is a proof or counterexample, it does not seem to fit in a method. When P is a 3SP property, P^{dual}, P^{co}, P^{up} etc are usually omitted. The fourth column says the place in the text where that 3SP problem can be found.

Approximation property (AP)	No	Every X separable contains a Y with FDD such that X/Y has FDD	7.1.c
AP^{dual}	No	"	7.1.d
Asplund	Yes	$RNP^{\,dual}$	4.11
Asplund (weak)	No	$C[0,1] \to D[0,1] \to c_0(\Gamma)$	5.6
Asplund + WCG	No	$c_0 \to JL \to l_2(\Gamma)$	4.19.e
{Asplund + WCG}^{dual}	Yes		4.19.d
ANP-I	Yes		5.16.a
ANP-II; ANP-III	?	Yes, when X/Y has ANP-I	5.16.b
ANP $w*$-k	No	Not stable by products	5.16.d
B_X lies in the range of a vector measure	No	$l_2 \to Z_2 \to l_2$	5.21.c
$(B_X*, w*)$ does not contain $[0,\omega_1]$	Yes		4.19.b
$(B_X*, w*)$ does not contain $\beta\mathbb{N}$	?	Not having l_∞ as a quotient	4.19.e 3.2.b
Banach-Saks (BS)	Yes	Alternate BS + reflexive	4.6.b

BS (Alternate)	Yes	lifting	4.6.c
BS (weak)	No	$Ker(q+i) \rightarrow l_1 \times S \rightarrow c_0$	4.6.a
p-BS, $1 < p < \infty$	No	$l_p \rightarrow Z_p \rightarrow l_p$	4.6.e
p-BS (weak)	No	$Ker(q+i) \rightarrow l_1 \times S \rightarrow c_0$	4.6.a
Basis	No	Every X separable containing c_0 contains Y with basis; X/Y has basis	7.2.c 7.2.d
" (boundedly complete)	No	$l_2 \rightarrow D \rightarrow W^* \rightarrow 0$	7.2.e
" (unconditional)	No	$l_2 \rightarrow Z_2 \rightarrow l_2$	7.2.f
B-convexity	Yes	{*not containing* l_1}^{up}	3.2.j
BD	?		6.11
Bounded approximation property (*BAP*)	No	Every X separable contains a Y with *FDD* such that X/Y has *FDD*	7.1.c
BAP^{dual}	No	"	7.1.d
(*C*) (of Pol)	Yes		4.18
C-convexity	Yes	*super*-{*not containing* c_0}	3.2.m
Cech	No	$B \rightarrow JT^* \rightarrow l_2(\Gamma)$	4.14
Cech^{dual}	Yes	*Separable* ^{co}	4.14.h
Cech^{co}	No	$Z \rightarrow Z^{**} \rightarrow B$ with $B^*=JT$	4.14.i
C^k-smooth norm, $k \geq 2$	No	$l_2 \rightarrow Z_2 \rightarrow l_2$	5.3.a
c_0	Yes	*separably injective*	3.2.c
c_0-saturated	Yes	incomparability	3.2.e
c_0 (not containing)	Yes	C_1	3.2.e
c_0 (not having quotients)	Yes	lifting	2.4.e
$C[0,1]$	?		3.5
$C[0,1]$ (not containing)	Yes		3.5.a
$C(K)$-space	No	$c_0 \rightarrow c_0(l_\infty, r_n) \rightarrow c_0(C(\beta \mathbb{N} - \mathbb{N}))$	3.5.b

Complemented in a dual	No	$PB \to B \oplus l_1(I) \to JT^*$	3.7.c
cotype q	No	$l_q \to Z_q \to l_q$	5.1.b
cotype (finite)	Yes	super-{not containing C_0}	6.2; 3.2.1
C_p, $1 \leq p \leq \infty$	Yes	lifting	6.1
Dieudonné	?		6.6
Dual space	No	$l_2 \to D \to W^*$	3.7.b
Dunford-Pettis (DP)	No	$Ker(q+i) \to l_1 \times S \to c_0$	6.6.f
DP hereditary (DP_h)	No	$Ker(q+i) \to l_1 \times S \to c_0$	6.6.f
$(DP_h)^{dual}$	Yes	$Schur^{dual} = (C_\infty)^{dual}$	6.6.g
DP_{hq}	Yes	$(DP_h)^{dual}$	6.6.i
DP order p, $1 < p < \infty$	?		6.6.h
DP^3	Yes		6.6.o
DP reciprocal	?		6.6
DP somewhat	Yes	incomparability	6.6.k
DP surjective	No	Not stable by products	6.6.m
Finite dimensional	Yes	lifting	2.4.i
Finite dimensional decomposition (FDD)	No	Every X separable contains Y with FDD such that X/Y has FDD	7.1.c
$(FDD)^{dual}$	No	"	7.1.d
Fréchet smooth norm	?		5.3
Gâteaux smooth norm	No	$C[0, 1] \to D[0, 1] \to c_0([0, 1])$	5.5.b
Gelfand-Phillips	No	$c_0 \to C(K) \to C(K \backslash \mathbb{N})$	6.8
Gordon-Lewis l.u.s.t.	No	$l_2 \to Z_2 \to l_2$	7.2.h
Grothendieck property	Yes	lifting	2.4.6; 6.6.0

Grothendieck's theorem	Yes	3-lemma	6.3
GSG	No	$C[0, 1] \to D[0, 1] \to c_0([0, 1])$	4.19
Hahn-Banach smooth	No	not stable by $\| \ \|_1$-products	1.9.h
Hereditarily.Indecomposable	?		2.8.
Hilbert	No	$l_2 \to Z_2 \to l_2$	3.1
Hilbert (twisted)	?		3.1
Hilbert (weak)	No	$l_2 \to Z_2 \to l_2$	3.1
Hilbert-Schmidt	Yes	Jarchow's factorization	6.4
Jayne-Namioka-Rogers	Yes		5.11.c
Kadec (-Klee) renorming	?	Yes, when X/Y has *LUR* renorming	5.20
K_p, $1 \le p < \infty$;	Yes	$C_{max\{2, p*\}}$	6.7.d
Krein-Milman	?	stable by products?	5.15
L (of Emmanuele)	?		6.11
l_p, $1 < p < \infty$	No	$l_p \to Z_p \to l_p$	3.2.b
l_p,(not containing)	Yes	incomparability	3.2.d
l_p-saturated, $1 \le p < \infty$	Yes	incomparability	3.2.d
$L_p(0,1)$, $1 < p < \infty$	No	$L_p \to Z_p \oplus L_p \to L_p$	3.4.a
$L_p(0,1)$ (not containing)	?		3.4.
$\mathscr{L}_p$, $p \neq 1, \infty$	No	$l_q \to Z_q \to l_q$	3.3.a
l_1	Yes	*projective*	3.2.c
$L_1(0,1)$	Yes	*dual injective*	3.4.b
$L_1(0,1)$ (not containing)	No	Not stable by products	3.4.c
$\mathscr{L}_1$	Yes	*dual injective*	3.3.b
l_∞	Yes	*injective*	3.2.c
l_∞ (not containing)	Yes		3.2.f

$\mathscr{L}_\infty$	Yes	*dual* $\mathscr{L}_1$	3.3.b
LUR renorming	Yes		5.10.g
Lyapunov	Yes		5.21.a
Mazur property	?		4.9
MIP	?		5.12
MIP^{dual}	Yes		5.12.e
Mazur rotations problem	?	Isometric problem: no	5.23
Measure compact	No	$C[0,1] \to D[0,1] \to c_0(\Gamma)$	5.9
Midpoint *LUR*	No	$c_0(\Theta\backslash\Lambda) \to C_0(\Theta) \to c_0(\Lambda)$	5.14.a
Never condensed	?		5.24
Octahedral norm (not)	Yes	*Not containing* l_1	5.19
Orlicz property	Yes	finite cotype; $0(X) \le 0(Y)\,0(X/Y)$	6.2
$id_X \in \Pi_{p,1}$	No	$l_p \to Z_p \to l_p$	6.2.c
Plichko-Valdivia (*PV*)	No	$c_0 \to JL \to l_2(\Gamma)$	4.12
PV^{co}	No	$Z \to Z^{**} \to Z^{**}/Z = B,\ (B^*=JT)$	
$PV^{\,dual}$	Yes	*Separable-by-Reflexive+Asplund*	4.13
PCP	Yes		4.17
Polish ball	Yes	*dual separable* + *PCP*	4.14
Polynomials are wsc	No	$Ker(q+i) \to l_1 \times S \to c_0$	4.20
Polynomial C_1	?		4.20
Polynomial V	?		4.20
Polynomial V^*	?		4.20
Polynomial Schur	?	Λ-space	4.20
PRI	No	$c_0 \to JL \to l_2(\Gamma)$	4.15
Q-reflexivity	?		4.20

Radon-Nikodym (*RNP*)	Yes	Jarchow's factorization	6.5
RNP (analytical)	?	Yes, when Y has *RNP*	6.5.c
RNP (near)	Yes	Jarchow's factorization	6.5.e
RNP (weak)	?		6.5.g
(*RNP* weak) ^{dual}	Yes	*Not containing l_1*	6.5.h
Reflexive-by-Separable	No	$c_0 \to JL \to l_2$	4.10
Reflexivity	Yes	exact functor	4.1
co-reflexivity	Yes	*Reflexive* ^{co}	4.3
complemented "	?		4.3
quasi-reflexivity	Yes	lifting	4.2
reflexivity somewhat	Yes	incomparability	4.4
superreflexivity	Yes	*Reflexive* ^{up}	4.5
Rough norm (not having)	Yes	*Asplund*	5.4.a
Schur	Yes	lifting	6.1
separable-by-reflexive	Yes		2.9.g
separability	Yes	lifting	2.4.h
Separately distinguished	No	$c_0 \to JL \to l_2(\Gamma)$	4.12.h
Separating polynomial	No	$l_2 \to Z_2 \to l_2$	4.20
σ-fragmentability	Yes		4.16
Skipped Blocking Decomposition	Yes		7.4
" boundedly complete	Yes	*PCP*	7.4.b
" reflexive	Yes	*Polish ball*	7.4.b
" shrinking	Yes	*separable^{dual}*	7.4.b
(not) stable ($X \neq X \times X$)	No	Not stable by products	3.6
Strictly convex norm	No	$c_0(\Theta \backslash \Lambda) \to C_0(\Theta) \to c_0(\Lambda)$	5.17

Szlenk index countable	Yes	$Sz(X) \leq Sz(Y) \, Sz(X/Y)$	5.13
type p	No	$l_p \to Z_p \to l_p$	5.1.b
type $< p$	Yes		5.1.c
(u)	No	$KerT_p \to X_p \to c_0$	6.9.b
U.M.D.	No		5.22
Uniform λ-property	No	Not stable by products	5.18.a
Uniformly convex	Yes	*super-reflexivity*	5.10.b
" (weak)	?		5.13
Unique predual	?		3.7
(V)	No	$KerT_p \to X_p \to c_0$	6.9.b
(V^*)	?		6.9
(w)	No	Not stable by products	6.10
WCG	No	$c_0 \to JL \to l_2(\Gamma)$	4.10
WCG^{dual}	?		4.10.g
WCG^{co}	No	$Z \to Z^{**} \to B$, with $B^*=JT$	4.10.h
WCD	No	$c_0 \to JL \to l_2(\Gamma)$	4.15
Weakly K-analytic	No	$c_0 \to JL \to l_2(\Gamma)$	4.16
W_p, $1<p<\infty$;	No	$l_{p*} \to Z_{p*} \to l_{p*}$	4.6
W_p weak $1\leq p<\infty$	No	$Ker(q+i) \to l_1 \times S \to c_0$	4.6.a
weak sequential completeness	Yes	lifting	4.7; 2.4.c
weak* sequentially compact dual ball	No	$c_0 \to C(K) \to C(K\backslash\mathbb{N})$	4.8
w^*-G_δ extreme points	Yes		5.8.b
WU	?		2.8
(X)	?		3.7

Bibliography

1   ADAM, E., BISTRÖM, P. AND KRIEGL, A., *Countably evaluating homomorphisms on real function algebras,* preprint 1995.

2   AKSOY, A.G AND KHAMSI, M.A., *Nonstandard methods in fixed point theory,* Springer Universitext 1990.

3   ALEXANDROV, G.A., *On the three-space problem for MLUR renormings of Banach spaces,* C.R. Acad. Bulgare Sci. 42 (1989), 17-20.

4   ALVAREZ, J.A., ALVAREZ, T. AND GONZÁLEZ, M., *The gap between subspaces and perturbation on non-semi-Fredholm operators,* Bull. Austral. Math. Soc. 45 (1992) 369-376.

5   ALVAREZ T. *A note on generalized tauberian operators.* Quaestiones Math. 10 (1987) 377-390

6   ALVAREZ T., GONZÁLEZ M. AND ONIEVA V.M., *A note on three-space Banach space ideals.* Arch. Math. 46 (1986) 169-170.

7   ALVAREZ T., GONZÁLEZ, M. AND ONIEVA, V.M., *Totally incomparable Banach spaces and three-space Banach space ideals.* Math. Nachr. 131 (1987) 83-88.

8   AMIR, D. AND LINDENSTRAUSS, J., *The structure of weakly compact sets in Banach spaces,* Ann. of Math. 88 (1968) 35-46.

9   ANANTHARAMAN, R. AND DIESTEL, J., *Sequences in the range of a vector measure,* Comment. Math. Prace. Math. 30 (1991) 221-235.

10 ARON, R. AND DINEEN, S., *Q-reflexive Banach spaces,* Rocky Mtn. J. Math. (to appear).

11 ARON, R. AND LOHMAN, R.H., *A geometric function determined by extreme points of the unit ball of a normed space,* Pacific J. Math. 127 (1987) 209-231.

12 ARRANZ, F., *Uncomplemented copies of C(K) inside C(K),* Extracta Math. 11 (1996) 412-413.

13 ASPLUND, E., *Fréchet differentiability of convex functions,* Acta Math. 121 (1968) 31-47.

14   BAERNSTEIN II, A., *On reflexivity and summability*, Studia Math. 42 (1972) 91-94.

15   BANACH, S., *Théorie des opérations linéaires*, Warszawa 1932.

16   BANACH, S. AND SAKS, S., *Sur la convergence forte dans les champs L_p*, Studia Math. 2 (1930) 51-57.

17   BANDYOPADHYAY, P., AND BASU, S., *On a new asymptotic norming property*, ISI Tech. Report 5, 1995.

18   BASU, S. AND RAO, T.S.S.R.K., *Some stability results for asymptotic norming properties of Banach spaces*, preprint 1996.

19   BATOR, E., *Remarks on completely continuous operators*, Bull. Polish Acad. Sci. 37 (1989) 409-413.

20   BEAUZAMY, B. AND LAPRESTÉ, J.T., *Modèles étalés des espaces de Banach*, Hermann, 1984.

21   BEHRENDS, E., *Points of symmetry of convex sets in the two dimensional complex space. A counterexample to D. Yost's problem,* Math. Ann. 290 (1991) 463-471.

22   BENYAMINI, Y., *An M-space which is not isomorphic to a C(K) space*, Israel J. Math. 28 (1977) 98-102.

23   BERGH, J. AND LÖFSTRÖM, J., *Interpolation spaces, an introduction*, Springer 1976.

24   BESSAGA, C. AND PELCZYNSKI, A., *Banach spaces not isomorphic to their cartesian squares* I, Bull. Acad. Polon. Sci. 8 (1960) 77-80.

25   BOMBAL, F., *Sobre algunas propiedades de espacios de Banach*, Rev. Real Acad. Ciencias Madrid 84 (1990) 83-116.

26   BOMBAL, F., *Distinguished subsets in vector sequence spaces*, Progress in Functional Analysis, K.D. Bierstedt, J. Bonet, J. Horvath and M. Maestre (eds), North-Holland Math. Studies 170 (1992) 293-306.

27   BOMBAL F., CEMBRANOS P. AND MENDOZA J., *On the Surjective Dunford-Pettis Property*. Math. Z. 204 (1990) 373-380.

28   BONET, J., DIEROLF, S. AND FERNÁNDEZ, C., *On the three-space problem for distinguished Fréchet spaces*, Bull. Soc. Roy. Liège 59 (1990) 301-306.

29   BOULDIN, R., *The instability of non semi-Fredholm operators under compact*

perturbations, J. Math. Anal. Appl. 87 (1982) 632-638.

30   BOURGAIN, J., *Dunford-Pettis operators on L_1*, Israel J. Math. 37 (1980) 34-47.

31   BOURGAIN, J., *l_∞/c_0 has no equivalent strictly convex norm*, Proc. Amer. Math. Soc. 78 (1980), 225-226.

32   BOURGAIN, J., *New classes of $\mathcal{L}_p$-spaces*, Lecture Notes in Math. 889, Springer 1981.

33   BOURGAIN, J., *Some remarks on Banach spaces in which martingale difference sequences are unconditional*, Arkiv der Math. 21 (1983) 163-168.

34   BOURGAIN, J., *New Banach space properties of the disc algebra and H^∞*, Acta Math. 152 (1984) 1-48.

35   BOURGAIN, J., *Real isomorphic Banach spaces need not be complex isomorphic*, Proc. Amer. Math. Soc. 96 (1986) 221-226.

36   BOURGAIN, J. AND DELBAEN, F. *A class of special $\mathcal{L}_\infty$-spaces,* Acta Math. 145 (1980) 155-176.

37   BOURGAIN, J., AND DIESTEL, J., *Limited operators and strict cosingularity*, Math. Nachr. 119 (1984) 55-58

38   BOURGAIN J. AND PISIER G., *A construction of $\mathcal{L}_\infty$-spaces and related Banach spaces*. Bol. Soc. Bras. Mat., 14 (1983) 109-123.

39   BOURGAIN, J. AND ROSENTHAL, H.P., *Geometrical implications of certain finite dimensional decompositions*, Bull. Soc. Math. Belgique 32 (1980) 57-82.

40   BOURGAIN, J, AND TALAGRAND, M., *Dans un espace réticulé, la propriété de Krein-Milman et celle de Radon-Nikodym sont équivalentes*, Proc. Amer. Math. Soc. 81 (1981) 93-96.

41   BROWN, A.L., *On the space of subspaces of a Banach space*, J. London Math. Soc. 5 (1972) 67-73.

42   BROWN, L. AND ITO, T., *Some non-quasi-reflexive spaces having unique isomorphic preduals*, Israel J. Math. 20 (1975) 321-325.

43   BUKHVALOV, A.V. AND DANILEVICH, A.A., *Boundary properties of analytic and harmonic functions with values in a Banach space*, Math. Zametki 31 (1982) 203-214; english translation: Math. Notes 31 (1982) 104-110.

44   BURKHOLDER, D.L., *A geometrical characterization of Banach spaces in which martingale difference sequences are unconditional*, Ann. of Prob. 9 (1981) 997-

1011.

45   CABELLO SÁNCHEZ, F., *The blackboard sessions 1996*.

46   CABELLO SÁNCHEZ, F., *Diez variaciones sobre un tema de Mazur*, Tesis, Univ. de Extremadura 1996.

47   CABELLO SÁNCHEZ, F., *Transitivity of M-spaces and Wood's conjecture*, Math. Proc. Cambridge Philos. Soc. 123 (1998) (to appear).

48   CABELLO SÁNCHEZ, F. AND CASTILLO, J.M.F., *Duality and twisted sums of Banach spaces*, Preprint 21 Univ. Extremadura, 1996; announced as *Report on twisted sums of Banach spaces*, Extracta Math. 11 (1996) 384-387.

49   CARNE, T.K., COLE, B. AND GAMELIN, T.W., *A uniform algebra of analytic functions on a Banach space*, Trans. Amer. Math. Soc. 314 (1989) 639-659.

50   CARRO, M.J., CERDÁ, J. AND SORIA, J., *Commutators in interpolation theory*, Ark. Math. 33 (1995) 199-216.

51   CARROLL, L., *The hunting of the Snark*, 1876.

52   CASSAZA, P.G. AND KALTON, N.J., *Notes on approximation properties in separable Banach spaces*, in *Geometry of Banach spaces*, Proc. Conf. Strobl, Austria 1989; P.F.X. Muller, W. Schachermayer (eds.) London Math. Soc. Lecture Notes 158 (1990) 49-63.

53   CASSAZA, P.G. AND SHURA, T., *Tsirelson's space*, Lecture Notes in Math. 1363, Springer 1989.

54   CASTILLO, J.M.F., *A variation on Schreier's space*, Riv. Mat. Univ. Parma, 2 (1993) 319-324

55   CASTILLO, J.M.F., *On Banach spaces X such that $L(L_p, X) = K(L_p, X)$*, Extracta Math. 10 (1995) 27-36.

56   CASTILLO, J.M.F., *On c_0 and l_1 saturated Banach spaces*, Preprint 12, Univ. Extremadura 1996.

57   CASTILLO, J.M.F., *Commutators and twisted sums*, Preprint 13 Univ. Extremadura 1996.

58   CASTILLO, J.M.F., *Snarked sums of Banach spaces*, Proceedings of the II Congress in Banach spaces, Badajoz, November 1996, Extracta Math. 12 (1997) (to appear).

59   CASTILLO, J.M.F. AND GONZÁLEZ, M. *An approach to Schreier space*, Extracta

Math. 6 (1991) 166-169.

60 CASTILLO, J.M.F. AND GONZÁLEZ, M., *The Dunford-Pettis property is not a three space property*, Israel J. Math. 81 (1993) 297-299.

61 CASTILLO, J.M.F. AND GONZÁLEZ, M., *Properties (V) and (u) are not three-space properties*, Glasgow Math. J. 36 (1994) 297-299.

62 CASTILLO, J.M.F. AND GONZÁLEZ, M., *New results on the Dunford-Pettis property*, Bull. London Math. Soc. 27 (1995) 599-605.

63 CASTILLO, J.M.F., GONZÁLEZ, M. AND YOST, D., *Twisted properties of Banach spaces*, preprint 1997.

64 CASTILLO, J.M.F. AND SÁNCHEZ, F., *Dunford-Pettis-like properties of continuous function vector spaces*, Rev. Mat. Univ. Complutense Madrid 6 (1993) 43-59.

65 CASTILLO, J.M.F. AND SÁNCHEZ, F., *Remarks on some basic properties of Tsirelson's space*, Note di Mat. 13 (1993) 117-122.

66 CASTILLO, J.M.F. AND SÁNCHEZ, F., *Remarks on the range of a vector measure*, Glasgow Math. J. 36 (1994) 157-161.

67 CASTILLO, J.M.F. AND SÁNCHEZ, F., *Weakly p-compact, p-Banach-Saks and super-reflexive Banach spaces*, J. Math. Anal. Appl. 185 (1993) 256-261.

68 CASTILLO, J.M.F. AND SIMOES, M.A., *On p-summable sequences in locally convex spaces*, Preprint 11, Univ. Extremadura 1994.

69 CHEN, D. AND LIN, B.L., *Ball separation properties in Banach spaces*, preprint 1995.

70 CIVIN, P. AND YOOD, B., *Quasi-reflexive spaces*, Proc. Amer. Math. Soc. (1957) 906-911.

71 CLARK, J.R., *Coreflexive and somewhat reflexive Banach spaces*, Proc. Amer. Math. Soc. 36 (1972) 421-427.

72 CLARKSON, J., *Uniformly convex spaces*, Trans. Amer. Math. Soc. (1936) 396-414.

73 COBAN, M. AND KENDEROV, P.S., *Dense Gateaux differentiability of the-sup norm in C(T) and the topological properties of T*, C. R. Acad. Bulgare Sci. 38 (1985) 1603-1604.

74 CORSON, H.H., *The weak topology of a Banach space*, Trans. Amer. Math. Soc. 101 (1961) 1-15.

75   CROSS, R.W., OSTROVSKII, M.I. AND SHEVCHIK V.V., *Operator ranges in Banach spaces I,* Math. Nachr. 173 (1995) 91-110.

76   CWIKEL, M., JAWERTH, B., MILMAN, M. AND ROCHBERG, R., *Differential estimates and commutators in interpolation theory,* in *Analysis at Urbana II,* Berkson E.R., Peck, N.T. and Uhl, J. (eds.), London Math. Soc. Lecture Notes Series 138 (1989) 170-220.

77   DAVIE, A.M., *The approximation problem for Banach spaces,* Bull. London Math. Soc. 5 (1973) 261-266.

78   DAVIS, W.J., FIGIEL, T., JOHNSON, W. AND PELCZYNSKI, A., *Factoring weakly compact operators,* J. Funct. Anal. 17 (1976) 311-327.

79   DE MARÍA, J.L., AND RODRÍGUEZ SALINAS, B., *The space l_∞/c_0 is not a Radon space,* Proc. Amer. Math. Soc. (1991) 1095-1100.

80   DEVILLE, R. AND GODEFROY, G., *Some applications of projective resolutions of identity,* Proc. London Math. Soc. 67 (1993) 183-199.

81   DEVILLE R., GODEFROY G. AND ZIZLER V., *The three space problem for smooth partitions of unity and C(K) spaces,* Math. Ann 288 (1990) 612-625.

82   DEVILLE, R. GODEFROY, G. AND ZIZLER, V., *Smoothness and renormings in Banach spaces,* Monographs and Surveys in Pure and Applied Math. 64, Longmann 1993.

83   DÍAZ, J.C., *On the three-space property: non-containment of l_p, $1 \leq p < \infty$, or c_0 subspaces,* Publ. Mat. 37 (1993) 127-132.

84   DÍAZ, J.C., DIEROLF, S., DOMANSKI, P. AND FERNÁNDEZ, C., *On the three space problem for dual Fréchet spaces,* Bull. Polish Acad. Sci. 40 (1992) 221-224.

85   DIEROLF, S., *Über Vererbbarkeitseigenschaften in topologischen Vektorräumen,* Dissertation, Universität München 1973.

86   DIEROLF, S., *On the three-space problem and the lifting of bounded sets,* Collect. Math 44 (1993) 81-89.

87   DIESTEL, J., *Geometry of Banach spaces,* Lecture Notes in Math. 485, Springer 1975.

88   DIESTEL, J., *A survey of results related to the Dunford-Pettis property,* In Amer. Math. Soc. Contemporary Mathematics, 2 (1980) 15–60.

89   DIESTEL, J., *Sequences and Series in Banach Spaces*, Graduate Texts in Math. 92, Springer-Verlag 1984.

90   DIESTEL, J., JARCHOW, H. AND TONGE, A., *Absolutely summing operators*, Cambridge studies in advanced mathematics 43, Cambridge Univ. Press 1995.

91   DIESTEL, J. AND SEIFERT, C.J., *An averaging property of the range of a vector measure*, Bull. Amer. Math. Soc. 82 (1976) 907-909.

92   DIESTEL, J. AND SEIFERT, C.J., *The Banach-Saks ideal I. Operators acting on* $C(\Omega)$, Comment. Math. Prace. Math. Special Vol. dedicated to W. Orlicz I (1978) 109-118.

93   DIESTEL, J. AND UHL JR., J.J., *Vector Measures*, Math. Surveys 15, Amer. Math. Soc., 1977.

94   DINEEN, S., *A Dvoretzky theorem for polynomials*, Proc. Amer. Math. Soc. 123 (1995) 2817-2821.

95   DIXMIER, J., *Étude sur les varietés et les opérateurs de Julia,* Bull. Soc. Math. France 77 (1944) 11-101.

96   DOMANSKI, P., *Local convexity of twisted sums*, in *Proceedings of the 12th Winter School on abstract analysis*, Srni, 1984; Suppl. Rend. Circolo Mat. di Palermo, 5 (1984) 13-31.

97   DOMANSKI, P., *Extensions and liftings of linear operators*, Poznan 1987.

98   DOMANSKI P., *On the projective LB-spaces*, Note di Mat. 12 (1992) 43-48.

99   DOWLING, P.N., *Representable operators and the Analytic Radon-Nikodym property in Banach spaces*, Proc. Roy. Irish Acad. A 85 (1985) 143-150.

100 DOWLING, P.N. AND EDGAR, G.A., *Some characterizations of the Analytic Radon-Nikodym property in Banach spaces*, J. Funct. Anal. 80 (1988) 349-357.

101 DREWNOWSKI, L., *Equivalence of Brooks-Jewet, Vitali-Hahn-Saks and Nikodym theorems*, Bull. Polish Acad. Sci. 20 (1972) 725-731.

102 DREWNOWSKI, L., *On minimally subspace-comparable F-spaces,* J. Funct. Anal. 26 (1977) 315-332.

103 DREWNOWSKI, L. AND EMMANUELE, G., *On Banach spaces with the Gelfand-Phillips property II*, Rend. Circ. Mat. Palermo 38 (1989) 377-391.

104 DREWNOWSKI, L. AND ROBERTS, J.W., *On Banach spaces containing a copy of* l_∞ *and some three-space properties*, unpublished.

105 DULST, D. VAN, *Characterizations of Banach spaces not containing l_1*, C.W.I. Tracts 59, Amsterdam 1989.

106 EDGAR G.A., *Measurability in a Banach space*, Indiana Univ. Math. J. 26 (1977) 663-677.

107 EDGAR G.A., *Measurability in a Banach space* II, Indiana Univ. Math. J. 22 (1979) 559-579.

108 EDGAR, G.A. AND WHEELER, F., *Topological properties of Banach spaces*, Pacific. J. Math. 115 (1984) 317-350.

109 ELTON, J. Thesis, Yale Univ.

110 EMMANUELE, G., *A dual characterization of Banach spaces not containing l_1*, Bull. Polish Acad. Sci. 34 (1986) 155-160.

111 EMMANUELE, G., *The (BD) property in $L_1(\mu, E)$*, Indiana Math. J., 36 (1987) 229-230.

112 EMMANUELE, G., *On the Dieudonné property*, Comment. Math. Prace. Math. 27 (1988) 253-256.

113 EMMANUELE, G., *Banach spaces in which the Dunford-Pettis sets are relatively compact*, Arch. Math. 58 (1992) 477-485.

114 ENFLO, P., *Banach space which can be given an equivalent uniformly convex norm*, Israel J. Math 13 (1972) 281-288.

115 ENFLO, P., *A counterexample to the approximation problem in Banach spaces*, Acta Math. 130 (1973) 309-317.

116 ENFLO P., LINDENSTRAUSS J. AND PISIER G., *On the "three space problem"*, Math. Scand. 36 (1975) 199-210.

117 FABIAN, M., *Projectional resolution of the identity sometimes implies weakly compact generating*, Bull. Polish Acad. Sci. (1990) 117-120.

118 FABIAN, M. AND GODEFROY, G., *The dual of every Asplund space admits a projectional resolution of the identity*, Studia Math 91 (1988) 141-151.

119 FIGIEL, T., unpublished manuscript.

120 FIGIEL T., GHOUSSOUB N. AND JOHNSON W.B., *On the structure of non-weakly compact operators on Banach lattices*, Math. Ann. 257 (1981) 317-334.

121 FIGIEL, T. AND JOHNSON, W.B., *The approximation property does not imply the*

bounded approximation property, Proc. Amer. Math. Soc. 41 (1973) 197-200.

122 FINET, C., *Duaux transfinis*, Math. Ann. 268 (1984) 293-304.

123 FLORET, K., *Weakly compact sets*, Lecture Notes in Math. 801, Springer 1980.

124 FREEDMAN, W., *Alternative Dunford-Pettis properties*, Doctoral dissertation, Univ. California, June 1995.

125 GARCÍA, R., *A note on the Dunford-Pettis property for quotients of C(K)*, Extracta Math. 10 (1995) 179-182.

126 GELFAND, I., *Abstrakte Funktionen und linearen Operatoren*, Math. Sb. 46 (1938) 235-284.

127 GHOUSSOUB, N. AND JOHNSON, W.B., *Counterexamples to several problems on the factorization of bounded linear operators*, Proc. Amer. Math. Soc. 92 (1984) 233-238.

128 GHOUSSOUB, N. AND MAUREY, B., *G_δ-embeddings in Hilbert spaces*, J. Funct. Anal. 61 (1985) 72-97.

129 GHOUSSOUB, N. AND MAUREY, B., *The asymptotic norming and Radon-Nikodym properties are equivalent in separable Banach spaces*, Proc. Amer. Math. Soc. 94 (1985) 665-671.

130 GIESY, D. P., *On a convexity condition in normed linear spaces*, Trans. Amer. Math. Soc. 125 (1966) 114-146.

131 GILES, J.R., GREGORY D.A, AND SIMS, B., *Characterization of normed linear spaces with Mazur's intersection property*, Bull. Austral. Math. Soc. 18 (1978) 105-123.

132 GODEFROY, G., *Compacts de Rosenthal,* Pacific J. Math. 91 (1980) 293-306.

133 GODEFROY, G., *Existence and uniqueness of isometric preduals: a survey*, in Amer. Math. Soc. Contemporary Math. 85 (1989) 131-193

134 GODEFROY, G. AND SAAB, P., *Quelques espaces de Banach ayant les propriétés (V) ou (V*) de A. Pelczynski*, C.R.A.S. Paris 303 (1986) 503-506.

135 GODEFROY, G. AND SAPHAR, P., *Three-space problems for the approximation properties*, Proc. Amer. Math. Soc. 105 (1989) 70-75.

136 GODEFROY, G., TROYANSKI, S., WHITFIELD, J. AND ZIZLER, V., *Three-space problem for locally uniform rotund renorming of Banach spaces*, Proc. Amer. Math. Soc. 94 (1985) 647-652.

137 GODUN, B.V. AND RAKOV, S.A., *Banach-Saks property and the problem of three spaces*, Mat. Zametki 31 (1982) 32-39; english transl. Math. Notes 31 (1982) 32-38.

138 GOLDBERG, S., *Unbounded linear operators*, McGraw-Hill, 1966.

139 GOLDMAN, M.A., *On the stability of the property of normal solvability of linear equations* (russian), Dokl. Acad. Nauk. SSSR 100 (1995), 201-204.

140 GONZÁLEZ, M., *Dual results of factorization for operators*, Ann. Acad. Sci. Fennicae, Ser. A, 18 (1993) 3-11.

141 GONZÁLEZ, M., *Remarks on Q-reflexive Banach spaces*, Proc. Roy. Irish. Acad. 96 (1996) 195-201.

142 GONZÁLEZ, M., GONZALO, R. AND JARAMILLO, J., *Separating polynomials on rearrangement invariant function spaces,* preprint.

143 GONZÁLEZ, M. AND GUTIÉRREZ, J., *Unconditionally converging polynomials on Banach spaces*, Math. Proc. Cambridge Philos. Soc. 117 (1995) 321-331.

144 GONZÁLEZ, M. AND MARTINÓN, A., *On incomparability of Banach spaces*, in *Functional Analysis and Operator Theory*, Banach Center Publications 30 (1994) 161-174.

145 GONZÁLEZ, M. AND ONIEVA, V.M., *On the instability of non-semi-Fredholm operators under compact perturbations*, J. Math. Anal. Appl. 114 (1986), 450-457.

146 GONZÁLEZ, M. AND ONIEVA, V.M., *On incomparability of Banach spaces*, Math. Z. 192 (1986) 581-585.

147 GONZÁLEZ, M. AND ONIEVA, V.M., *Ideal functions and operator ideals*, Publ. Sem. Mat. García Galdeano, Serie II, sec. 1, n° 15 (1988)

148 GONZÁLEZ, M. AND ONIEVA, V.M., *Lifting results for sequences in Banach spaces*, Math. Proc. Cambridge Philos. Soc. 105 (1989) 117-121.

149 GONZÁLEZ, M. AND ONIEVA, V.M., *El problema de los tres espacios*, In *Homenaje al Prof. Nacere Hayek Calil* (1990) 155-162.

150 GONZÁLEZ, M. AND ONIEVA, V., *Characterizations of tauberian operators and other semigroups of operators*, Proc. Amer. Math. Soc. 108 (1990) 399-405.

151 GONZALO, R. AND JARAMILLO, J.A., *Smoothness and estimates of sequences in Banach spaces*, Israel J. Math. 89 (1995) 321-341.

152 GONZALO, R. AND JARAMILLO, J.A., *Compact polynomials between Banach spaces*, Proc. Roy. Irish Acad. 95 (1995) 1-14.

153 GOUWELEEUW, J.M., *A characterization of vector measures with convex range*, Proc. London Math. Soc. 70 (1995) 336-362.

154 GOWERS, W.T., *A solution to Banach's hyperplane problem*, Bull. London Math. Soc. 26 (1994) 523-530.

155 GOWERS, W.T., *A new dichotomy for Banach spaces*, Geometric and Functional Analysis 6 (1996) 1083-1093.

156 GOWERS, W.T. AND MAUREY, B., *The unconditional basic sequence problem*, J. Amer. Math. Soc. 6 (1993) 851-874.

157 GOWERS, W.T. AND MAUREY, B., *Banach spaces with small spaces of operators*, Math. Ann. (to appear).

158 GROTHENDIECK, A., *Sur le applications faiblement compactes d'espaces du type C(K)*, Canad. J. Math. 5 (1953) 129-173.

159 GUTIÉRREZ, J.M., JARAMILLO, J.A. AND LLAVONA, J.G., *Polynomials and Geometry of Banach spaces*, Extracta Math. 10 (1995) 79-114.

160 HAGLER, J., *A counterexample to several questions about Banach spaces*, Studia Math. 60 (1977) 289-308.

161 HAGLER, J. AND SULLIVAN, F., *Smoothness and weak sequential compactness*, Proc. Amer. Math. Soc. 78 (1980) 497-503

162 HARMAND, P., WERNER, D. AND WERNER, W., *M-ideals in Banach spaces and Banach algebras*, Lecture Notes in Math. 1547, Springer 1993

163 HARTE, R., *Invertibility, singularity and Joseph L. Taylor*, Proc. Roy. Irish Acad. 81 (1981) 71-79.

164 HAYDON, R., *The three-space problem for strictly convex renormings and the quotient problem for Fréchet smooth renormings*, Séminaire d'Initiation à l'Analyse, exp. 1, Publ. Math. Univ. Pierre et Marie Curie 107, Paris 1995.

165 HAYDON, R., *Trees in renorming theory*, preprint 1995.

166 HE, S., *The ω property in dual spaces*, J. Central China Normal Univ. Natur. Sci., 25 (1991) 159-161.

167 HEINRICH, S., *Ultraproducts in Banach space theory*, J. Reine Angew Math. 313 (1980) 72-104.

168 HERMAN, R. AND WHITLEY, R., *An example concerning reflexivity*, Studia Math. 28 (1967) 289-294.

169 HILTON, P.J., AND STAMMBACH, U., *A course in homological algebra*, Graduate Texts in Math. 4, Springer 1970.

170 HU, Z. AND LIN, BOR-LUH, *On the asymptotic norming property of Banach spaces*, in *Function spaces*, Lecture notes in Pure and Appl. Math. 136 (1991) 195-210, Marcel Dekker.

171 HU, Z. AND LIN, BOR-LUH, *Smoothness and asymptotic norming properties in Banach spaces*, Bull. Austral. Math. Soc. 45 (1992) 285-296.

172 HU, Z. AND LIN, BOR-LUH, *Three-space problem for the asymptotic norming property of Banach spaces*, Contemporary Math., Amer. Math. Soc. 144 (1993) 135-140.

173 HU, Z. AND LIN, BOR-LUH, *RNP and CPCP in Lebesgue-Bochner function spaces*, Illinois J. Math. 37 (1993) 329-347.

174 JAMES, R.C., *A non-reflexive Banach space isometric with its second conjugate*, Proc. Nat. Acad. Sci. USA 37 (1951) 174-177.

175 JAMES, R.C., *Uniformly non square Banach spaces*, Ann. of Math. 80 (1964), 542-550.

176 JAMES, R.C., *Some self-dual properties of normed linear spaces*, Ann. of Math. Studies 69 (1972), 159-175.

177 JAMES, R.C., AND HO, A., *The asymptotic norming and Radon-Nikodym properties for Banach spaces*, Ark. Math. 19 (1981) 53-70.

178 JAMESON, G.J.O., *Topology and normed spaces*, Chapmann and Hall, London 1974.

179 JANZ, R., *Perturbation of Banach spaces*, unpublished preprint Univ. Konstaz, 1987.

180 JARAMILLO, J. AND PRIETO, A., Weak polynomial convergence on a Banach space, Proc. Amer. Math. Soc. 118 (1993) 463-468.

181 JARAMILLO, J.A., PRIETO, A., AND ZALDUENDO, I., *The bidual of the space of polynomials on a Banach space*, Math. Proc. Cambridge Philos. Soc. (to appear); a report can be found in Extracta Math. 9 (1994) 124-131.

182 JARCHOW, H., *Locally convex spaces*, B.G. Teubner 1981.

183 JARCHOW, H., *On Hilbert-Schmidt spaces,* Suppl. Rend. Circolo Mat. di Palermo, II-2 (1981) 153-160.

184 JARCHOW, H., *The three space problem and ideals of operators*, Math. Nachr. 119 (1984) 121-128

185 JARCHOW, H., *Remarks on compactness of operators defined on* L_p , Note di Mat. 11 (1991) 225-230.

186 JAWERTH, B., ROCHBERG, R. AND WEISS, G., *Commutator and other second order estimates in real interpolation theory,* Ark. Math. 24 (1986) 191-219.

187 JAYNE, J.E., NAMIOKA, I. AND ROGERS, C.A., *σ-fragmentable Banach spaces,* Mathematika 39 (1992) 161-188.

188 JAYNE, J.E., NAMIOKA, I. AND ROGERS, C.A., *Topological properties of Banach spaces*, Proc. London Math. Soc. 66 (1993) 651-672.

189 JIMÉNEZ, M. AND MORENO, J.P., *Renorming Banach spaces with the Mazur intersection property,* J. Funct. Anal. 144 (1997) 486-504.

190 JOHNSON, W.B., *Operators into* L_p *which factor through* l_p, J. London Math. Soc. 14 (1976) 333-339.

191 JOHNSON, W.B., *On quotients of* L_p *which are quotients of* l_p , Compo. Math. 34 (1977), 69–89.

192 JOHNSON, W.B., *Extensions of* c_0, preprint 1997.

193 JOHNSON, W.B. AND LINDENSTRAUSS, J., *Some remarks on weakly compactly generated Banach spaces,* Israel J. Math. 17 (1974) 219-230.

194 JOHNSON, W.B., LINDENSTRAUSS, J. AND SCHECHTMAN, G., *On the relation between several notions of unconditional structure,* Israel J. Math. 37 (1980) 120-129.

195 JOHNSON, W.B. AND ROSENTHAL, H.P., *On w*-basic sequences and their applications to the study of Banach spaces,* Studia Math. 43 (1972) 77-92

196 JOHNSON. W.B. AND ZIPPIN, M., *Subspaces and quotient spaces of* $(\Sigma G_n)_p$ *and* $(\Sigma G_n)_0$, Israel J. Math. 17 (1974) 50-55.

197 JOHNSON. W.B. AND ZIPPIN, M., *Extension of operators from weak*-closed subspaces of* l_1 *into C(K) spaces,* Studia Math. 117 (1995) 43-55.

198 KADETS, M.I., *Note on the gap between subspaces,* Funct. Anal. Appl. 9 (1975), 156-157.

199 KADETS, V. AND SCHECHTMANN, G, *The Lyapunov theorem for l_p-valued measures*, St. Peterburg Math. J. 4 (1993) 961-966.

200 KADETS, V. AND VLADIMIRSKAYA, O.I., *Three-space problem for Lyapunov theorem on vector measures*, preprint 1996.

201 KAIBKHANOV, K.E., *On Banach spaces that are not isomorphic to their Cartesian squares*, Mat. Zametki 57 (1995) 534-541; english transl. Math. Notes 57 (1995) 369-374.

202 KAKUTANI, S., *Weak convergence in uniformly convex spaces*, Math. Institute Osaka Imperial Univ. (1938) 165-167.

203 KALTON, N.J., *The three-space problem for locally bounded F-spaces*, Compo. Math. 37 (1978) 243-276.

204 KALTON, N.J., *Convexity, type and the three space problem*, Studia Math. 69 (1981) 247-287.

205 KALTON, N.J., *Curves with zero derivatives in F-spaces*, Glasgow Math. J. 22 (1981) 19-29.

206 KALTON, N.J., *A symplectic Banach space with no Lagrangian subspaces*, Trans. Amer. Math. Soc. 273 (1982) 385-392.

207 KALTON, N.J., *Nonlinear commutators in interpolation theory*, Mem. Amer. Math. Soc. 385 (1988).

208 KALTON, N.J., *Differentials of complex interpolation processes for Köthe function spaces*, Trans. Amer. Math. Soc. 333 (1992) 479-529.

209 KALTON, N., *An elementary example of a Banach space not isomorphic to its complex conjugate*, Canad. Math. Bull. 38 (1995) 218-222.

210 KALTON, N.J., *The basic sequence problem*, Studia Math. 116 (1995) 167-187.

211 KALTON, N.J., *The existence of primitives for continuous functions in a quasi-Banach space*, Atti sem. Mat. e Fisico dell'Università di Modena 44 (1996) 113-118.

212 KALTON, N.J. AND PECK, N.T. *Quotients of $L_p(0, 1)$, for $0 \leq p < 1$*, Studia Math. 64 (1979) 65-75.

213 KALTON, N.J. AND PECK, N.T., *Twisted sums of sequence spaces and the three space problem*, Trans. Amer. Math. Soc. 255 (1979) 1-30.

214 KALTON, N.; PECK, N.T. AND ROBERTS, *An F-space sampler*, London Math.

Soc. Lecture Note Series 89, Cambridge Univ. Press 1984.

215 KALTON, N.J. AND PELCZYNSKI, A., *Kernels of surjections from $\mathcal{L}_1$-spaces with an applications to Sidon sets*, Math. Ann. (to appear).

216 KALTON, N.J., AND ROBERTS, J.W., *Uniformly exhaustive submeasures and nearly additive set functions*, Trans. Amer. Math. Soc. 278 (1983) 803-816.

217 KAPPELER, T., *Banach spaces with the condition of Mazur*, Math. Z. 191 (1986) 623-631.

218 KATO, T., *Perturbation theory for linear operators*, Springer-Verlag 1976.

219 KAUFMAN, R., PETRAKIS, M., RIDDLE, L.H. AND UHL JR., J.J., *Nearly representable operators*, Trans. Amer. Math. Soc. 312 (1989) 315-333.

220 KELLEY, J.L., *General topology*, Graduate Texts in Math. 27, Springer 1955.

221 KISLYAKOV, S.V., *Spaces with "small" annihilators*, J. Soviet Math. 16 (1981) 1181-1184.

222 KNAUST, H. AND ODELL, E. , *On c_0-sequences in Banach spaces*, Israel J. Math. 67 (1989), 153-169.

223 KNAUST, H. AND ODELL, E., *Weakly null sequences with upper l_p-estimates*, in Functional Analysis, Lecture Notes in Math. 1470, Springer 1991, 85-107.

224 KORDULA, V. AND MULLER, V., *Vasilescu-Martinelli formula for operators in Banach spaces*, Studia Math. 113 (1995) 127-139.

225 KÖTHE, G., *Topological vector spaces I*, Springer 1969.

226 KREIN, M. AND SMULLIAN, V., *On regularly convex sets in the space conjugate to a Banach space*, Ann. of Math. 41 (1940) 556-583.

227 KRIVINE, J.L. AND MAUREY, B., *Espaces de Banach stables*, Israel J. Math. 39 (1981) 273-295.

228 LABUSCHAGNE, L.E., *On the instability of non-semi-Fredholm closed operators under compact perturbations with applications to ordinary differential operators*. Proc. Edinburg Math. Soc. 109A (1988), 97-108.

229 LANCIEN, G., *Indices de dentabilité et indice de Szlenk*, C.R.A.S. 314 (1992) 905-910.

230 LANCIEN, G., *Théorie de l'indice et problèmes de renormage en géométrie des espaces de Banach*, Thesis, Paris 1992

231 LANCIEN, G., *Dentabilité indices and locally uniformly convex renormings*, Rocky Mtn. J. Math. 23 (1993) 635-647.

232 LANCIEN, G., *On uniformly convex and uniformly Kadec-Klee renormings*, Serdica Math. J. 21 (1995) 1-18.

233 LANG, S., *Real and functional analysis*, Graduate Texts in Math. 142, Springer 1993.

234 LARMAN, D.G. AND PHELPS, R.R., *Gâteaux differentiability of convex functions on Banach spaces*, J. London Math. Soc. 20 (1979) 115-127.

235 LEMBERG, H., *Nouvelle demonstration d'un théoreme de J.L. Krivine sur la finie representation de l_p dans un espace de Banach*, Israel J. Math. 39 (1981) 341-348.

236 LEUNG D., *Uniform convergence of operators and Grothendieck spaces with the Dunford-Pettis Property*, Math. Z. 197 (1988) 21-32 .

237 LEUNG, D., *On Banach spaces with Mazur's property*, Glasgow Math. J. 33 (1991) 51-54.

238 LEUNG D., *Banach Spaces with Property (w)*, Glasgow Math. J. 35 (1993) 207-217.

239 LEUNG D., *On the Weak Dunford-Pettis Property*, Arch. Math.

240 LIMA, A., *Intersection properties of balls and subspaces in Banach spaces*, Trans. Amer. Math. Soc. 227 (1977) 1-62.

241 LIMA, A., AND YOST, D., *Absolutely Chebyshev subspaces,* in: S. Fitzpatrick and J. Giles (eds.), Workshop/Miniconference Funct. Analysis/Optimization, Canberra 1988; Proc. Cent. Math. Analysis Australian Nat. Univ. 20 (1988) 116-127.

242 LIN, BOR-LUH AND ZHANG, W., *Three-space property of Kadec-Klee renorming in Banach spaces*, preprint.

243 LINDENSTRAUSS, J., *On a certain subspace of l_1*, Bull. Polish Acad. Sci. 12 (1964) 539-542.

244 LINDENSTRAUSS, J., *On James' paper "Separable conjugate spaces,"* Israel J. math. 9 (1971) 279-284.

245 LINDENSTRAUSS, J., *Weakly compact sets - their topological properties and the Banach spaces they generate*, Ann. Math. Studies 69 (1972) 235-273.

246 LINDENSTRAUSS, J. AND PELCZYNSKI, A., *Contributions to the theory of classical Banach spaces*, J. Funct. Anal. 8 (1971) 225-249.

247 LINDENSTRAUSS, J. AND ROSENTHAL, H.P., *The $\mathcal{L}_p$- spaces*, Israel J. of Math. 7 (1969) 325-349.

248 LINDENSTRAUSS, J. AND TZAFRIRI, L., *Classical Banach spaces I, sequence spaces*, Springer 1977.

249 LINDENSTRAUSS, J. AND TZAFRIRI, L., *Classical Banach spaces II, function spaces*, Springer 1979.

250 LOHMAN, R.H., *A note on Banach spaces containing l_1*, Canad. Math. Bull. 19 (1976) 365-367

251 LOHMAN, R.H., *The λ-function in Banach spaces*, Amer. Math. Soc. Contemporary Math. 85 (1989) 345-354.

252 LOHMAN, R.H., *Aspects of the uniform λ-property*, Note di Mat. 12 (1992) 157-165.

253 LOZANOVSKII, G.Y., *On some Banach lattices*, Siberian Math. J. 10 (1969) 419-430.

254 LUSKY, W., *A note on rotations in separable Banach spaces*, Studia Math. 65 (1979) 239-242.

255 LUSKY, W., *A note on Banach spaces containing c_0 or c*, J. Func. Anal. 62 (1985) 1-7.

256 MARKUS, A.S., *On some properties of linear operators connected with the notion of gap* (russian), Kishinev Univ. Uchen. Zap. 39 (1959), 265-272.

257 MASCIONI, V., *On weak cotype and weak type in Banach spaces*, Note di Mat. 8 (1988) 67-110.

258 MAUREY, B. AND PISIER, G., *Séries de variables aléatoires vectorielles indépendants et géométrie des espaces de Banach*, Studia Math. 58 (1976) 45-90.

259 MAZUR, S., *Über schwache Konvergenz in den Räumen L^p*, Studia Math. 4 (1933) 128-133.

260 MESHKOV, V.Z., *Smoothness properties in Banach spaces*, Studia Math. 63 (1978) 111-123.

261 MOLTÓ, A., *On a theorem of Sobczyk*, Bull. Austral. Math. Soc. 43 (1991) 123-

130.

262 MOLTÓ, A., ORIHUELA, J. AND TROYANSKI, S., *Locally uniformly rotund renorming and fragmentability*, Proc. London Math. Soc. (to appear).

262* MÜLLER, V., *The splitting spectrum differs from the Taylor spectrum*, Studia Math. (to appear).

263 MUSIAL, K., *The weak Radon Nikodym property in Banach spaces*, Studia Math. 64 (1979) 151-173.

264 MUSIAL, K., *The weak Radon Nikodym property in conjugate Banach spaces*, in *Banach space theory and its applications*, Lecture Notes in Math. 991 Springer 1981, 182-186.

265 NAMIOKA, I. AND PHELPS R.R., *Banach spaces which are Asplund spaces*, Duke Math. J. 42 (1975) 735-750.

266 NICULESCU, C.P., *Absolute continuity in Banach space theory*, Rev. Roumaine Math. Pures Appl. 24 (1979) 413-422.

267 NISHIURA, T. AND WATERMANN, D., *Reflexivity and summability*, Studia Math. 23 (1963) 53-57.

268 ODELL, E., *On quotients of Banach spaces having shrinking unconditional bases*, Illinois J. Math. 36 (1992) 681-695.

269 ONIEVA, V.M., *Notes on Banach space ideals*, Math. Nachr. 126 (1986), 27-33.

270 ORLICZ, W., *Beiträge zur Theorie der Orthogonalentwicklungen II*, Studia Math. 11 (1929) 241-255.

271 OSTROVSKII, M.I., *Banach-Saks properties, injectivity and gaps between subspaces of a Banach space*, Teor. Funktsii, Funktsion. Anal. i Prilozhen 44 (1985) 69-78; english transl. J. Soviet Math. 48 (1990) 299-306.

272 OSTROVSKII, M.I., *Three space problem for the weak Banach-Saks property*, Math. Zametki 38 (1985) 721-725.

273 OSTROVSKII, M.I., *On properties of the opening and related closeness characterizations of Banach spaces*. Amer. Math. Soc. Transl. 136 (1987), 109-119.

274 OSTROVSKII, M.I., *Properties of Banach spaces stable and unstable for the gap*, Sibirsk Math. Zh. 28 (1987) 182-184; english transl. Siberian Math. J. 28 (1987) 140-142.

275 OSTROVSKII, M.I., *Topologies on the set of all subspaces of a Banach space and related questions of Banach space geometry*, Quaestiones Math. 17 (1994) 259-319.

276 OSTROVSKII, M.I. AND PLICHKO, A.N., *Banach-Saks properties and the three space problem* (russian) In *Operators in function spaces and problems in function theory*, Kiev, Naukova Dumka (1987) 96-105.

277 PARTINGTON, J.R., *On the Banach-Saks property*, Math. Proc. Cambridge. Philos. Soc. 82 (1977) 369–374.

278 PATTERSON, W.M., *Complemented c_0-subspaces of a non-separable $C(K)$-space*, Canad. Math. Bull. 36 (1993) 351-357.

279 PAYÁ, R. AND RODRÍGUEZ-PALACIOS, A., *Banach spaces which are semi-L-summands in their biduals*, Math. Ann. 289 (1991) 529-542.

280 PAYÁ, R. AND YOST, D., *The two-ball property: transitivity and examples*, Mathematika 35 (1988) 190-197.

281 PELCZYNSKI, A., *Banach spaces on which every unconditionally converging operator is weakly compact*, Bull. Polish Acad. Sci. 10 (1962) 641-648.

282 PELCZYNSKI, A., *Linear extensions, linear averagings, and their applications to linear topological classification of spaces of continuous functions*, Dissertationes Math. 58 (1968).

283 PERROTT, J.C., *Transfinite duals of Banach spaces and ergodic super-properties equivalent to super-reflexivity*, Quart. J. Math. Oxford 30 (1979) 99-11.

284 PHILLIPS, R.S., *On linear transformations*, Trans. Amer. Math. Soc. 48 (1940) 516-541.

285 PIETSCH, A., *Operator Ideals*, North Holland 1980.

286 PISIER, G., *Sur les espaces qui ne contiennent pas de l_1^n uniformément*, Séminaire Maurey-Schwartz 1973/74, exposé VII.

287 PISIER, G., *B-convexity, superreflexivity and the three space problem*, Publ. Aarhaus Univ. (1975) 151-175.

288 PISIER, G., *Some applications of the complex interpolation method to Banach lattices*, J. Analyse Math. 35 (1979) 264-281

289 PISIER, G., *Holomorphic semigroups and the geometry of Banach spaces*, Ann. of Math. 115 (1982) 375-392.

290 PISIER, G., *Counterexamples to a conjecture of Grothendieck*, Acta Math. 151 (1983) 181-208.

291 PISIER, G., *Factorization of linear operators and geometry of Banach spaces*, Regional conf. series in Math. 60, Amer. Math. Soc. 1984.

292 PISIER, G., *Probabilistic methods in the geometry of Banach spaces*, in Probability and analysis, Springer Lecture Notes in Math. 1206, 1984.

293 PISIER, G., *Sur les operaterus p-sommants et p-radonnifiants pour p < 1*, Astérisque 131 (1985) 163-174.

294 PISIER, G., *Weak Hilbert spaces*, Proc. London Math. Soc. 56 (1988) 547-579.

295 PLICHKO, A.N., *Some properties of Johnson-Lindenstrauss space*, Funktional Anal. i Prilocen 15 (1981) 88-89; english transl. Functional Anal. and Appl. 15 (1981) 149-150.

296 POL, R., *On a question of H.H. Corson and some related problems*, Fundamenta Math. 109 (1980) 143-154.

297 POPOV, M.M., *On integrability in F-spaces*, Studia Math. 110 (1994) 205-220.

298 PREISS, D., PHELPS, R.R. AND NAMIOKA, I., *Smooth Banach spaces, weak Asplund spaces and monotone or usco mappings*, Israel J. Math. 72 (1990) 257-279.

299 QIYUAN, N., *Schauder basis determining properties*, Acta Math. Scientia 12 (1992) 89-97.

300 RAKOV, S.A., *Banach spaces in which a theorem of Orlicz is not true*, Mat. Zametki 14 (1973) 101-106; english transl. Math. Notes 14 (1973) 613-616.

301 RAKOV, S.A., *C-convexity and the "problem of three spaces"*, Soviet Math. Dokl. 17 (1976) 721-724

302 RAKOV, S.A., *Ultraproducts and the "three space problem"*, Funct. Anal. Appl. 11 (1977) 236-237.

303 RAKOV, S.A., *Banach-Saks property of a Banach space*, Math. Zametki 6 (1979) 823-834; english transl. Math. Notes 26 (1979) 909-916.

304 READ, C.J., *Different forms of the approximation property*, unpublished manuscript 1989.

305 RIBARSKA, N.K., *Three-space property for σ-fragmentability*, Mathematika (to appear).

306 RIBE, M., *Examples for the nonlocally convex three space problem*, Proc. Amer. Math. Soc. 237 (1979) 351-355.

307 ROCHBERG, R. AND WEISS, G., *Derivatives of analytic families of Banach spaces*, Ann. of Math. 118 (1983) 315-347.

308 RODRÍGUEZ-PALACIOS, A., *Proper semi-L-embedded complex subspaces*, Studia Math. 106 (1993) 197-202.

309 ROSENTHAL, H.P., *On totally incomparable Banach spaces*, J. Funct. Anal. 4 (1969) 167-175

310 ROSENTHAL, H.P., *On injective Banach spaces and the spaces C(S)*, Bull. Amer. Math. Soc. 75 (1969) 824-828.

311 ROSENTHAL, H.P., *On relatively disjoint families of measures, with some applications to Banach space theory*, Studia Math. 37 (1970) 13-36.

312 ROSENTHAL, H.P., *On L_p-spaces*, Ann. of Math. (1973) 344-373.

313 ROSENTHAL, H.P., *A characterization of Banach spaces not containing l_1*, Proc. Nat. Acad. Sci. (USA) 71 (1974) 2411-2413

314 ROSENTHAL, H.P., *Embeddings of L_1 in L_1*, Contemp. Math 26, Amer. Math. Soc. 1984, 335-349.

315 ROSENTHAL, H.P., *Weak* polish Banach spaces*, J. Funct. Anal. 76 (1988) 267-316.

316 RYAN, R., *The Dunford-Pettis property and projective tensor products*, Bull. Polish Acad. Sci. 35 (1987) 785-792.

317 SAAB, E. AND SAAB, P. *On stability problems for some properties in Banach spaces*, in *Function spaces*, K. Jarosz ed.; Marcel Dekker Lecture Notes in Pure and Appl. Math. 136 (1992) 367-394.

318 SAAB, E. AND SAAB, P., *Extensions of some classes of operators and applications*, Rocky Mount. J. Math. 23 (1993) 319-337.

319 SAKAI, S., Math. Reviews 27, 5 (1964) 5107.

320 SCHACHERMAYER, W., *For a Banach space isomorphic to its square the Radon Nikodym property and the Krein Milman property are equivalent*, Studia Math. 81 (1985) 329-339.

321 SCHLUMPRECHT, T., *Limited sets in Banach spaces*, Ph. D. Thesis Munich (1988).

322 SCHLUMPRECHT, T., *An arbitrarily distortable Banach space*, Israel J. Math. 76 (1991) 81-95.

323 SCHLUMPRECHT, T., *Limited sets in C(K) spaces and examples concerning the Gelfand-Phillips property*, Math. Nachr. 157 (1992) 51-64

324 SCHLUCHTERMANN, G. AND WHEELER, R.F., *On strongly WCG Banach spaces*, Math. Z. 199 (1988) 387-398.

325 SCHREIER, J., *Ein Gegenbeispiel zur Theorie der swachen Konvergenz*, Studia Math. 2 (1930) 58-62.

326 SEMADENI, Z., *Banach spaces non-isomorphic to their Cartesian square*, Bull. Polish Acad. Sci 8 (1960) 81-84.

327 SIMS, B., *"Ultra" techniques in Banach space theory*, Queen's papers in Pure and Applied Math. 60, Queen's University, Kingston, Ontario, Canada 1982.

328 SIMS, B. AND YOST, D., *Linear Hahn-Banach extension operators*, Proc. Edinburgh Math. Soc. 32 (1989) 53-57.

329 SINGER, I., *Bases in Banach spaces II*, Springer 1981.

330 STEGALL, C., *The Radon-Nikodym property in conjugate Banach spaces*. Trans. Amer. Math. Soc. 206 (1975) 213-223.

331 STEGALL, C., *The Radon-Nikodym property in conjugate Banach spaces II*. Trans. Amer. Math. Soc. 264 (1981) 507-519.

332 STEHLE, S., *On the hereditary Dunford-Pettis Property*, Thesis Michigan 1989.

333 STRAEULI,, E.W., *On extension and lifting of operators*, Dissertation, Univ. Zürich 1985.

334 STRAEULI, E., *On Hahn-Banach extensions for certain operator ideals*, Arch. Math. 47 (1986) 49-54.

335 SZAREK, S.J., *A superreflexive Banach space which does not admit complex structure*, Proc. Amer. Math. Soc. 97 (1986) 437-444.

336 SZAREK, S., *A Banach space without a basis which has the bounded approximation property*, Acta Math. 159 (1987) 81-98.

337 SZLENK, W., *The nonexistence of a separable reflexive Banach space universal for all separable reflexive Banach spaces*, Studia Math. 30 (1968) 53-61.

338 TALAGRAND, M., *Espaces de Banach faiblement K-analytiques*, Ann. of Math.

110 (1979) 407-438.

339 TALAGRAND, M., *Deux exemples de fonctions convexes*, C.R.A.S. 288 (1979) 461-464.

340 TALAGRAND, M., *Un nouveau C(K) qui possède la propriété de Grothendieck*, Israel J. Math. 37 (1980) 181-191.

341 TALAGRAND, M., *Rénormages de quelques C(K)*, Israel J. Math. 54 (1986) 327-334.

342 TALAGRAND, M., *The three space problem for L_1*, J. Amer. Math. Soc. 3 (1990) 9-29.

343 TALAGRAND, M., *Cotype of operators from C(K)*, Invent. Math. 107 (1992) 1-40.

344 TALAGRAND, M., *Cotype and (q, 1)-summing norm in a Banach space*, Invent. Math. 110 (1992) 545-556.

345 TAYLOR, J.L., *A joint spectrum for several commuting operators*, J. Funct. Anal. 6 (1970) 172-191.

346 TAYLOR, J.L., *Analytic functional calculus for several commuting operators*, Acta Math. 125 (1970) 1-38.

347 TERENZI, P., *Basic sequences and associated sequences of functionals*, J. London Math. Soc. 50 (1994) 187-198.

348 TROYANSKI, S., *On locally uniformly convex and differentiable norms in certain nonseparable Banach spaces*, Studia Math. 37 (1971) 173-180.

349 VALDIVIA, M., *On a class of Banach spaces*, Studia Math. 60 (1977) 11-13.

350 VALDIVIA, M., *Banach spaces X with X** separable*, Israel J. Math. 59 (1987) 107-111.

351 VALDIVIA, M., *Banach spaces of polynomials without copies of l_1*, Proc. Amer. Math. Soc. 123 (1995) 3143-3150.

352 VASAK, L., *On one generalization of weakly compactly generated Banach spaces*, Studia Math. 70 (1981) 11-19.

353 VASILESCU, F.H., *A Martinelli type formula for the analytic functional calculus*, Rev. Roumaine Math. Pures Appl. 23 (1978) 1587-1605.

354 VEECH, W.A., *A short proof of Sobczyk's theorem*, Proc. Amer. Math. Soc. 28 (1971) 627-628.

355 VOGT, D., *Lectures on projective spectra of (DF) spaces*, Lectures held in the Functional Analysis Seminar, Dusseldorf/Wuppertal, Jan-Feb 1987, unpublished manuscript.

356 WEBB, J.H., *Exact sequences and quasi-reflexivity*, Mat. Coll. Cape Town 6 (1970/71) 28-32.

357 WHEELER, R.F., *The retraction property, CCC property, and Dunford-Pettis-Phillips property for Banach spaces*, Lecture Notes in Math. Springer 945 (1982) 252-262.

358 WILANSKY, A., *Mazur spaces*, Internat J. Math. Sci. 4 (1981) 39-53.

359 YANG, K., *A note on reflexive Banach spaces*, Proc. Amer. Math. Soc. 18 (1967) 859-861.

360 YANG, K., *The generalized Fredholm operators*, Trans. Amer. Math. Soc. 216 (1976) 313-326.

361 YANG, K., *The reflexive dimension of an R-space*, Acta Math. Hungarica 35 (1980) 249-255.

362 YOST, D., *Irreducible convex sets,* Mathematika 38 (1991) 134-155.

363 YOST , D., *Asplund spaces for beginners*, Acta Univ. Carolinae, 34 (1993) 159-177.

364 YOST, D., Explanations on a coffe table of the contents of his talk at the Northwest European Analysis Seminar *Reducible convex sets* (Lyon 26-28 May 1995).

365 YOST, D. *The e-mail sessions, 1996-97.*

366 YOST, D., *On the Johnson-Lindenstrauss space*, Proceedings of the II Congress in Banach spaces, Badajoz, November 1996, Extracta Math. 12 (1997).

367 ZAIDENBERG, M.G., KREIN, S.G., KUCHMENT, P.A. AND PANKOV, A.A., *Banach bundles and linear operators*, Russian Math. Surveys 30 (1975) 115-175.

368 ZHANG, W., *Some geometrical and topological properties in Banach spaces*, Thesis, Iowa Univ. 1991.125

Subject index